EUREKA STOCKADE

A ferocious and bloody battle

EUREKA STOCKADE
A ferocious and bloody battle

Gregory Blake

BIG SKY PUBLISHING
www.bigskypublishing.com.au

Big Sky Publishing Pty Ltd
PO Box 303, Newport, NSW 2106, Australia
Phone: 1300 364 611
Fax: (61 2) 9918 2396
Email: info@bigskypublishing.com.au
Web: www.bigskypublishing.com.au

Cover design and typesetting: Think Productions

National Library of Australia Cataloguing-in-Publication entry (pbk.)
Author: Blake, Gregory, 1955-
Title: Eureka Stockade : a ferocious and bloody battle / Gregory Blake.
ISBN: 9781922132048 (pbk.)
Notes: Includes bibliographical references.
Subjects: Eureka Stockade (Ballarat, Vic.)--History. Victoria--History--1851-1891.
Dewey Number: 994.57031

National Library of Australia Cataloguing-in-Publication entry (ebook)
Author: Blake, Gregory, 1955-
Title: Eureka Stockade [electronic resource] : a ferocious and bloody battle / Gregory Blake.
ISBN: 9781922132055 (eBook)
Notes: Includes bibliographical references.
Subjects: Eureka Stockade (Ballarat, Vic.)--History. Victoria--History--1851-1891.
Dewey Number: 994.57031

CONTENTS

This book is dedicated to:

Albert Long 1939-2007.

Teacher, colleague, Australian history enthusiast and a great mate.

ACKNOWLEDGEMENTS

A narrative such as this could not exist without contributions by many people. It is therefore appropriate to acknowledge here those who assisted in my quest to discover the actual story of the fight for the Eureka Stockade.

Thanks must go in particular to military historian and good friend Peter Williams, for his encyclopaedic knowledge of the military milieu, never tiring eye for detail, razor sharp critique, and consistent encouragement to keep at the task. In a similar manner James Larkin and Paul Johnson provided valuable critique of early drafts. Specific details of military life during the mid nineteenth century would have remained mysteries if not for the knowledgeable contributions of Lance Caldwell, Gordon Findlay MacKinlay, Tim Pickles, and William Curtis.

Investigating the importance of the part played by the Americans, particularly the Californians, at Eureka would not have been possible without the timely advice of Robert Burke of Antioch California. The leads Robert provided opened up sources in the United States that were to be of inestimable value. Likewise, the kindly assistance provided by Rick Sherman of the Californian Genealogical Society in Oakland, California, and Pat Keats at the library of the Society of Californian Pioneers in San Francisco, proved to be of immense assistance. Appreciation must also be recorded for the consideration shown by the staff at the San Francisco National Historical Park, J. Porter Shaw Library.

The genuine interest shown in the project by John Lewis of the Sovereign Hill Museums Association, Ballarat, was of great assistance. The kindness shown to me by Jan Croggan, Senior Historian of that Association, and Roger Trudgeon, Curator and Deputy Director of the Gold Museum in Ballarat, was to prove invaluable. It was a pleasure to spend time on the telephone with Mary Akers, tapping into her intimate knowledge of Ballarat during the gold rush era. The personal anecdotes of Anne Lewis, who as a child played among the old tracks and mines of Black Hill, also added to my appreciation of the setting of the events at Eureka.

Catherine Green, Archivist at the Australasian College of Surgeons, provided friendly, prompt advice and service when I was hunting for forensic details of wounds caused by black powder weapons. Peter Bennett must also be acknowledged for the research work he did at the British National Archives. Finally, it would be remiss not to recognise Anthony Vidot and Phyllis Hall, without whose linguistic skills the meaning of the frequent digressions by Rafaello Carboni into languages other than English, both modern and archaic, would have remained forever obscure. To each and every one of these people go my heartfelt thanks, for without you this history would not have been possible.

INTRODUCTION

The ground was damp underfoot and a night chill hung in the air as a marching column of soldiers silently halted behind a low hill.[1] Somewhere close by, out to their front and hidden by the hill and the pre-dawn darkness, was their foe. It was just before dawn on Sunday 3 December 1854.

Without a word being spoken, the soldiers briskly moved into long-practiced formations. To the front went the skirmishers, careful to maintain the prescribed distance between them as they established their line. A little further back others closed up into more compact ranks, prepared to wait their moment in the coming affray. As the soldiers settled, they loosened the latches on their cartridge boxes for ease of access when the action began. For the same reason, the more experienced among them tucked loose cartridges into the waistline of their trousers. Shadowy shapes passed to the right and left of the infantry as mounted police and soldiers moved out to each flank. Officers on horseback trotted up and down ensuring all was as it should be. The lines stood still as they awaited the order that they knew must come. Despite the pre-dawn darkness a magistrate accompanying the soldiers could see a flag fluttering somewhere close ahead.[2]

As the sky lightened in anticipation of the new day's sun just beneath the eastern horizon, the order was given. Skirmishers began to move forward, each pace they took regulated as far as possible not to throw their formation into disorder. Extending their line, they passed around to the right of and partially over the hill to their front, then descended into a slight gully. As they did, silhouettes of the trees to the east slowly took on form. Somewhere hidden in the gloom beneath those trees, armed insurgents awaited them.

The soldiers peered into the dark, searching in vain for some sign of their adversaries. The line edged forward. It was deathly quiet.

With no warning a flame flashed out from the shadows. A lead ball cut through the air, making a sound akin to that of a mosquito. The ball struck home. Michael Roney, a private in the 40th Regiment, jerked back, killed instantly by a shot through the head.[3]

Other balls flew among the soldiers, one zipping dangerously close by an officer and a magistrate. More whistled through the ranks. The soldiers fingered their firelocks, no doubt wishing that they could return the fire, but their discipline held and not one of them did. They had but a moment to wait. The bugle's harsh notes shrilled 'Commence Firing', and a thunderous roar erupted as, with a sound like a file drawn across the teeth of a saw, several score military muskets were discharged.[4]

So began the battle for the Eureka Stockade. It was to be a savage but brief affair, lasting only 20 minutes. At its end more than 40 insurgents lay dead and at least 18 soldiers had fallen, two being killed outright, while several later died of their wounds. This was not a grand clash of massed battalions, batteries or charging squadrons. In the words of one of the military participants, it was in the military sense 'really nothing more than a trifling matter'.[5] Yet it was a battle nonetheless, and one that was intense, ferocious and sanguine while it lasted.

This narrative is a military history, an account of the battle at Eureka. It is a detailed account of a violent moment, when courage, terror, and carnage joined to create a legend which has since left an indelible imprint upon Australian history.

For too long our collective memory of the battle for the Eureka Stockade has been dictated by a script. The story line is well known and never changes. Innocent gold miners protesting against the harsh regime of a tyrannical government are set upon by hundreds of bloodthirsty soldiers and police. No warning is given, and the diggers, lacking arms and taken completely by surprise, are routed in a few brief minutes. A fearsome massacre then occurs as the military and police lose all control and run amuck, visiting murder and desolation on any unfortunate they could catch. Such is the legend. Such is the myth.

That this should be the case comes as no surprise. The many retellings of the story can be characterised by their brevity and chronic lack of rigorous historiography. In many cases the Eureka legend has been preferred above the facts to such an extent that the myth has become the story. Partisan wordsmiths began plying their trade within days of the battle and have not relented since. Such an approach to telling the story of the Eureka Stockade does an immense disservice to both history and the memory of those who fought and died there on that fateful morning.

In the wake of such consistently poor descriptions, it could well be argued that the whole event has become so hopelessly entangled within its own legend that it is impossible to unravel what occurred. Yet, as this narrative will reveal, such a gloomy assessment is far from the case.

In *The Face of Battle*, his seminal study of men in combat, John Keegan begins his account of the battle of Waterloo with a quote from the Duke of Wellington. Responding to requests to describe Waterloo in terms that could enable an historical narrative of the battle to be written, the Duke replied that the:

> history of the battle is not unlike the history of a ball! Some individuals may recollect all the little events of which the great result is the battle lost or won; but no individual can recollect the order in which, or the exact moment at which, they occurred, which makes all the difference to their value or importance.[6]

Such is the case with descriptions of the battle for the Eureka Stockade.

Despite there being a great deal of material available that enables us to construct a coherent narrative of the battle at Eureka, no one source can reveal all that we wish to know. The real challenge comes, as Wellington observed, when one wants to make sense from what one has discovered. This becomes especially so when we begin to unravel the tangle of half-truth and blatant mythology that has until now sufficed to portray the battle. Where, then, does one turn to begin to tell the authentic story of what happened at Eureka at dawn on that Sunday morning in December 1854?

The most obvious place to begin is the carefully worded official government reports of the day. The methodical military descriptions of what eventuated at Eureka and the dry statistical record of logistics, manpower, timings and expenditure, may appear on the surface to be somewhat uninspiring reading. However, such material provides us with a solid foundation from which to begin to construct a coherent understanding of the event.

Having established our base we then need to build on it, adding the body, soul and passion of the personal experience without which any description of human conflict would become essentially meaningless. For this we turn to the recollections, anecdotes and occasional diatribes of those most intimately involved in the battle. Luckily one does not have to look too far to find sources for such material. Rafaello Carboni's much-eulogised account *The Eureka Stockade, The Consequence of Some Pirates Wanting on Quarter Deck a Rebellion*, is an obvious place to start.

Carboni, a member of the insurgents' inner council, was not inside the stockade during the battle, but did claim to witness the conflict. He wrote his account in the year following Eureka, when his passion was still aflame and his memory fresh. Even though he is shamelessly bigoted in its opinions of participants, frequently linguistically pompous, and irredeemably self-serving, Carboni's account is compelling. His description of the battle occupies only a small part of his book; nevertheless he provides us with an invaluable resource that more often than not stands the test when compared with the recollections of others.

In the same manner journal entries, such as those of Thomas Pierson and shopkeeper Samuel Lazarus, are of great value. Often written only days after the event, these accounts offer us precious details that, when viewed in the military context of events, provide valuable clues as to what most probably did or did not occur. Letters from those who attacked and defended the stockade such as Captain Charles Pasley and Alfred Madocks, written in the weeks following the battle, are equally revealing. Contemporary newspaper articles, both foreign and domestic, even when written by quite obviously partisan correspondents, help enormously to piece together the jigsaw puzzle of what took place. Other resources, such as a note scrawled in the margin of a book by a defender of the stockade, also add to our overall appreciation of what occurred.

The many second-hand accounts of Eureka should not be dismissed. Even though those accounts are often flawed, they frequently provide details that corroborate the accounts of people much nearer to the event. The voluminous transcripts of the Eureka State Trials and court depositions are another rich source of first-hand material related to the fighting. In these transcripts the men who attacked the stockade recount at first hand, under oath and presumably as accurately as they can, their movements and actions during the conflict. When attempting to construct a military narrative, these accounts provide a great deal of detail on the battle.

Finally there are the numerous later reminiscences of those who defended or attacked the stockade, as well as those who watched from a distance. These are mostly accounts from by then quite elderly men and women. Despite in some cases being separated in time from the events by half a century or more, they provide us with a rich well spring from which we can draw and compare with what we know from other sources closer in time to the event.

Having uncovered the sources that will enable us to tell the human story, we must above all allow that story to tell itself. This is an important consideration if we wish to understand the event in the context of the times in which it occurred. It is easy to succumb to anachronistic judgements of what occurred at Eureka, as indeed have many who have told the story in the decades since. In Harper Lee's wonderful novel *To Kill a Mocking Bird*, Atticus Finch, a small town southern lawyer in the United States gives his young daughter some sage advice when he tells her that to get to know someone, you must first 'climb into his skin and walk around in it'. Sadly for our understanding of the battle at Eureka, there have been few modern tellers of the tale who have heeded this advice.

Without attempting to identify with those on both sides at Eureka, we cannot hope to appreciate fully what happened there. Unfortunately, this lack of empathy, especially in regard to the military mores of the era, has been most evident in many of the commentaries on the battle. The essential military illiteracy of the writers of many of these accounts has led to much nonsense being suggested and subsequently perpetuated as fact. It is from such blather that our collective memory of what happened at Eureka has been formed, and the heroes and villains of our national mythology subsequently determined.

As has been already pointed out, this is the history of a battle. To appreciate a battle, an understanding of the participants is required. With Eureka, it is also essential that we understand the military milieu of the era in which the battle occurred. This is quite specifically the paternalistic, brutally disciplined, culture of the British Army of 1854. Understanding this, we can begin to appreciate what was expected from soldiers and officers of the era, and their most likely reactions when under extreme duress.

We must also know something of the police who were at Eureka, distancing ourselves as much as possible from the one and half centuries of prejudice and vilification that have been heaped upon them.

Finally, there are the Eureka insurgents, the men who stood behind the slab barricades and for a time exchanged shot for shot and blow for blow with the soldiers and police. What made those ordinary men cast aside their inhibitions, take up arms and bring them to that fateful place? How were they armed and organised, what indeed were their intentions?

Most importantly we must understand the military technology of 1854, a fundamentally important aspect of the fight that has been all but ignored in every account. Only by appreciating all of these aspects can we hope to begin developing an informed appreciation of what happened at Eureka.

This history investigates each of these issues. As is often the case, the answers that emerge prompt even more questions. It is the answers to many of those later questions that force us to reassess everything we thought we knew about the battle for the Eureka Stockade.

What were the actual intentions of the Eureka insurgents? They appear not to have been as pacific as the legend insists. Where were the soldiers actually marching to in the pre-dawn darkness that morning? It becomes apparent that it must have been elsewhere than the currently accepted site for the stockade. Were the defenders really caught completely by surprise? Apparently not, given the intense fire encountered by the soldiers when they first showed themselves in front of the stockade.

If the firing from the stockade was haphazard, why did some of the soldiers waver, and why did it subsequently take them about ten minutes to cross the 150 yards (135 metres) of open ground to the stockade? What does this tell us about the supposedly poorly armed defenders of the stockade of legend? Who among those defenders was most likely to have been capable of producing such fire?

Why were the Americans, the one group there capable of delivering such fire, all but written out of the history of Eureka? In the same way, the decisive role played by the police in winning the victory has been consistently ignored. What does this say about our perceptions of who is and who is not permitted to play significant roles in our collective historical mythology? How many actually were wounded or died as a result of Eureka? Many more than admitted by both sides at the time. Even so, there was no gratuitous massacre at Eureka, an emotive pejorative much abused at the time and in a great many subsequent retellings of the story.

This history will prove conclusively that the battle for the Eureka Stockade was much harder fought than has ever been acknowledged. It was certainly not a simple civil disturbance, an event to be derided as of no importance. Nor was Eureka anything like a riot, a term chosen in the immediate aftermath by the colonial government and establishment press, to deliberately demean its importance. Eureka was much more. It was a full-blooded military engagement, albeit brief, but without doubt savage and hard fought.

Embark with me now upon a journey into a legendary moment in Australia's past. It was a moment when the allure of gold fired the unbounded aspirations of the common people. A moment when those aspirations challenged the entrenched social and political order, a bloody conflict erupted, and an Australian legend was born.

Chapter 1
TO PIERCE THE TYRANT'S HEART

When, many years later, Charles Derius Ferguson wrote about his experiences during the Eureka uprising, he recalled attending a meeting of his fellow American miners at the Adelphi Theatre. He remembered that some of the men there had been keen to join what he called the 'digger army'.[1] However, there were many present, including Ferguson, who declined to do so, even though he, and no doubt others like him, threw in their lot with the insurgents a short time later. It was, however, obvious from the tenor of the debate among the American miners at the meeting that there was a belief that an army of sorts had indeed been formed from the most disaffected elements among the Ballarat miners.

As events would later confirm, Ferguson's allusion to a 'digger army' was an overly complimentary description of the armed muster that had occurred. Yet it was a fact that many men had formed themselves into units and agreed to accept a degree of discipline. For several days armed companies had been openly drilling, a commander-in-chief had been elected, arms, ammunition, saddles and horses forcibly requisitioned, and a defensive structure erected. It was obvious to men like Ferguson that the purpose of the disaffected miners' muster could not have been anything other than martial in intent.

What made ordinary men willingly put aside their peaceful pursuits, take up arms, organise themselves and parade in such a manner? Why did they do so when they would have known that their actions could provoke a potentially deadly response from the governing authorities? These are fundamentally important questions the answers to which explain what lay at the very core of the Eureka uprising and why it took on the character that it did.

On the afternoon of Thursday 30 November 1854, several thousand miners gathered under a flagpole erected at Bakery Hill in Ballarat. Nailed atop the flagpole was a large blue and white Southern Cross flag. The flag, consisting of five eight-pointed white stars unified by a white cross on a dark blue field, was a simple yet strikingly effective design.

As the flag fluttered in the hot breeze of the late afternoon, the miners milled about. They were very angry. That morning the mounted police (often called troopers in that era) had gone onto the Gravel Pit diggings, not too far from the government camp, with drawn swords. They were hunting, in their rough and arbitrary manner, for any miner without a licence, a regular occurrence and the bane of everyone working the goldfields. Greeted with a shower of abuse, rocks and bottles, they had beaten a hasty retreat. A short while later they returned, supported by soldiers with fixed bayonets. More abuse, physical scuffles and some shots followed in the largest, most confrontational, and ultimately final of the infamous 'digger hunts' on the Ballarat goldfields. The riot at the Gravel Pit diggings was the catalyst for the release of long pent up fury and fear among the miners, triggering the armed confrontation three days later at Eureka.

It would be a mistake to imagine that unrest on the goldfields of the early 1850s was unique to Ballarat. There had been serious disturbances on other diggings in Victoria and New South Wales. In January 1853, several hundred miners took up arms to press their grievances at Sofala, on the Turon in New South Wales. Like their brethren at Ballarat just over a year later, the Bendigo miners had teetered precariously on the edge of armed rebellion during the later part of 1853.[2] In the same way, in 1854 a thousand gold miners at Castlemaine had rushed the government camp there to redress a wrong they felt had been done.[3] Yet unlike Ballarat, no armed insurgencies occurred and no blood was shed at any of these places. Why was this so?

"THE JOES!"

Known as 'Joes' the mounted police conducted frequently over zealous and sometimes brutal 'digger hunts' for miners who did not have licences. These hunts, and the humiliation inflicted on those who were apprehended, were a major source of discontent amongst miners. (*The Affair at Eureka*, Spielvogel).

Serious confrontation had been brewing for some time between the forces of the crown and the miner population on the Ballarat diggings. The persistently unreasonable and coercive manner in which the miner's licence laws were enforced in an environment of limited economic return for very hard physical labour was certainly a major aggravating factor. An even greater irritant was the widespread and growing perception among many miners at Ballarat that the very people entrusted with ensuring the laws were administered, the police and the courts, were irredeemably corrupt and self-serving.

This loss of faith in the integrity of the institutions of the law had steadily worked towards breaking down the unwritten social contract that normally existed between those who governed and those who were governed. A repressed rage simmered in the hearts of many men on the Ballarat goldfields. It was a rage only waiting for a spark to transform itself into violent fury. That spark came on the night of 6 October, when Scotsman James Scobie was murdered near the Eureka Hotel.

Scobie, along with his friend Peter Martin, had been involved in a drunken altercation with James Bentley, the hotel proprietor. Soon after, Scobie was murdered. When his body was found, he had been struck in the head. Bentley was blamed for the murder and held by a great many on the diggings to be unquestionably guilty of the crime. Scobie's murder set into motion a series of events that would ultimately result in open insurrection.

Bentley's trial was conducted at Ballarat and presided over by officials far too closely associated with the accused. He was acquitted. No clearer example of the despicable level to which those who administered the law on the Ballarat diggings had sunk could have been given to the miners, who had taken a keen interest in the trial. Regarded by them as a farce and a travesty of justice, the results of Bentley's trial brought thousands of angry men to the Eureka Hotel to protest.

In the resulting riot the hotel was first looted, then it and an adjacent bowling alley destroyed by fire. It was certainly an anarchic outburst, but one that signalled quite clearly the demise of the social restraints that had, up until then, kept the seething anger of the miners in check. The subsequent arrest of seven men for participating in the riot, on the most tenuous grounds, added further to the sense of outrage. Courts at Ballarat found four of these men not guilty, the remaining three were sent to Melbourne for trial and subsequently imprisoned. After that it simply did not matter that Bentley was retried in Melbourne and sentenced to three years for manslaughter. Whatever tenuous legitimacy the guardians of the law had enjoyed at Ballarat had been dealt a mortal blow.

In the ensuing climate, any heavy-handed act by the authorities was sure to be seen in its worst light. A tactful, considerate and tempered approach was called for. Such had been the response of the authorities at Sofala, Bendigo and Castlemaine. This would not be the case at Ballarat, where the powers of the police had been recognised as being absolute.[4]

In an example of the arbitrary powers exercised by some police, Johannes Gregorious, the crippled servant of Father Patrick Smyth, one of the two Catholic priests assigned to the Ballarat diggings, was arrested. Gregorious had been going to visit a sick man on 10 October when he was apprehended by a police trooper and arrested for not having a miner's licence. It did not seem to matter to the arresting officer that Gregorious was exempt from the requirement to possess a licence. His prosecution proceeded in the face of an extremely angry response from the Catholic community on the diggings, especially the Irish. For those outraged by the treatment of Gregorious it appeared that, once again, the role of the law to protect the weak and defenceless had been turned on its head.

It was in this environment that the more astute among the miners began to organise a political response to the challenges that faced them. Mass meetings were called, attended by thousands. These monster meetings, as the press dubbed them, were not unusual occurrences on the Victorian goldfields. Similar meetings had been held at Chewton, Bendigo and Castlemaine, and would continue to be held whenever there were matters of public concern to be decided. What made the meetings at Ballarat different was that they occurred in a climate of mutual distrust and naked antagonism between the goldfield authorities and the general population on the diggings. It would take an administration possessed of the wisdom of Solomon to calm such passions. Unfortunately, such wisdom was absent at Ballarat.

Lawlessness and anarchy certainly seemed to be threatening the Ballarat diggings. Yet for the colonial government of Victoria, there appeared to be an even more sinister threat lurking behind a façade of apparent mass hooliganism. The government had a genuine fear that there was a conspiratorial democratic political agenda in the hearts of a good many disaffected people at Ballarat. Such fears were confirmed when the Ballarat Reform League, established to represent the miners' grievances, adopted a political charter remarkably modern and democratic in its aspirations.

This charter was accepted as the guiding manifesto of the League at a mass meeting on 11 November, which 10,000 attended. Among its articles, the charter recognised the people, not the crown, as the source of ultimate sovereignty, and that it was the right of the people to remove any power that tyrannised them. There was also a not-so-veiled threat to separate the colony from Britain unless equal laws and rights were granted to all in what the charter referred to as a 'free community'.[5] Even though the Charter met with calculated indifference from the colonial government of Victoria, the dire threat to the social and political status quo posed by it would not have been lost on them.[6]

In an attempt to secure the release of the three Eureka Hotel riot prisoners who remained in custody, the Reform League despatched a delegation to Sir Charles Hotham, Governor of Victoria. In a sad example of the gulf that existed between the rigid social hierarchies and expectations of Hotham's world and the egalitarian spirit that had taken root among the mining population, he took exception to the use of the word *demand* by the delegates, and refused to see them. At the same time, mischievous newspaper reports claiming that the delegates had been arrested incensed many at Ballarat, who were ever willing to believe such rumours. Nevertheless, despite passions among the miners running hot, the Ballarat diggings remained remarkably free from any further outbreaks of violence. That was about to change.

On 28 November Captain Henry Wise arrived at the Ballarat diggings with a company of the 40th Regiment. The men had come up the 56 miles (90 kilometres) from the port of Geelong to Ballarat in carts, but Wise halted the carts just short of the diggings and dismounted his men. Ordering the soldiers to fix bayonets, load their muskets and prepare loose cartridges in their cartridge boxes, he marched them ostentatiously through the diggings to the government camp. This act may have been a deliberately provocative gesture or nothing more than bravado, but it was to have dire consequences later that evening, when a company of the 12th Regiment arrived on the diggings.

Like Wise's company of the 40th, the men of the 12th arrived on carts. Unlike Wise, the officer commanding did not dismount his men. Nor did he order them to fix bayonets or load their muskets. In fact, it appears that many of the muskets were not even being carried at the time, but were in the carts. Picking one's way through the diggings in the dead of night was a notoriously tedious procedure and the progress of the carts was slow. When passing close by the Eureka diggings, a place notorious for the misbehaviour of the Irishmen who dominated the claims there, the last two carts were set upon by a throng of assailants.

Chaos ensued as the carts were overturned and a civilian driver and two soldiers badly beaten. Tumbling out, the men in the leading carts formed up, but could do very little to intervene in the shambles that engulfed their comrades. Some time during the fray a number of shots were fired. One hit John Egan, a drummer boy of the 12th, in the thigh, while another hit the owner of a local store.[7] It was never determined who fired the shots.

Hearing the commotion, mounted police issued from the camp with swords drawn and rescued the battered and humiliated soldiers. Some accounts claim that this attack was prompted by a belief among some of the miners that the soldiers were bringing artillery onto the diggings. The very fear that such an action was possible underscored the collapse of trust between the miners and the authorities at Ballarat. Nonetheless, the attack was roundly condemned from all quarters, including many of the leading reformers, as a distinctly *un-British* thing to have occurred. Ironically, as we shall discover, it was in fact a very British thing in the circumstances.

Another monster meeting followed on 29 November at Bakery Hill, a low but prominent feature within sight of the government camp. Under the watchful eye of the police and soldiers, the massed gathering vented its anger, which was by now at fever pitch. Numerous motions condemning the authorities were carried and some licences were burned. If a situation required tact and forbearance this was it. Unfortunately, there was to be neither at Ballarat.

In an act of such myopic insensitivity that it has been claimed by some to have been deliberately designed to push the miners into open rebellion, a large scale 'digger hunt' was ordered for 30 November. At the time the Resident Gold Fields Commissioner at Ballarat, Robert Rede, justified his actions as a means of testing the mood of the miners, and claimed that he had received orders from the highest authority.[8] This hunt had the most profound consequences.

The very hot and blustery conditions on 30 November matched perfectly the explosive mood on the diggings. Troopers with drawn swords were the first to venture out, riding into the Gravel Pits near the government camp. The miners there responded with abuse and then a shower of improvised missiles. Unable to contain the situation, the troopers sent for assistance. This very quickly came in the form of infantry with fixed bayonets and loaded muskets. The crowd of miners at the Gravel Pits swelled in size. More missiles were thrown, more abuse hurled.

Rede, who was not a cowardly man, rode among the angry crowd in an attempt to either overawe or placate them. He remonstrated for while without success, then, surrendering any hope of prevailing by persuasion, began to read the Riot Act. Accounts vary on how effectively Rede did this. It seemed to some that he had not finished and to others that he had no sooner finished when the soldiers, bayonets held before them, and mounted police, swords drawn, advanced on the mass of miners.

A riot followed in which the infuriated crowd, refusing to cower, threw rocks, chunks of hard clay, pieces of wood and anything else that came to hand. A few struck out at the police, injuring several of them. They responded in kind, striking down several rioters. Then several shots were fired. Some accounts said that the soldiers fired a volley over the heads of the crowd; others that scattered shots came at random from both sides. Sweeping all before them with levelled bayonets, the Redcoats advanced to clear the Gravel Pits. The troopers rode forward on the flanks, taking several prisoners.

Word of the riot at the Gravel Pits raced across the diggings. Rightly or wrongly, news spread that the soldiers had fired on the miners. The fury and indignation were palpable. Even more compelling for many was that their own government appeared to have abandoned all restraint against them. For many miners this was the last straw, and they resolved that to remedy such a state of affairs required radical direct action, if only so that they could defend themselves against further attack.

That evening a mass of miners gathered once again on Bakery Hill. Many were armed this time, and kneeling together they swore an oath to defend each other beneath their Southern Cross flag. Then they began to organise themselves into armed companies. The next day they commenced erecting a stockade. The die had been cast at the Gravel Pits on 30 November, and what up until then had been an angry yet still mostly civil protest movement slid inexorably into an unambiguous armed insurrection.

How had it come to this? Why did so many on the Ballarat diggings feel their treatment at the hands of the authorities to be so offensive that they were prepared to take up arms and fight to redress their grievances? What was the essential spark that ignited the flame of insurrection within the hearts of the miners and resulted in the conflagration at Eureka?

Much has been written regarding what motivated the miners at Eureka. Some claim they were fighting for democracy, and that 'Australian democracy was born at Eureka', and proudly assert that the 'glorious constitution now enjoyed was cradled at Eureka'.[9] Such notions are bitterly rejected by others.[10] Eureka has been defined in Marxist terms as a struggle by the working class against their oppressors.[11] Anarchists see Eureka as an example of direct democratic action taken by a group of the self-employed who formed a mutually supportive collective.[12]

Some claim it was the beginnings of a fight for national independence and an Australian republic.[13] Another interpretation is that Eureka was a rebellion by greedy, self-centred and self-serving gold miners who didn't want to pay a tax.[14] Access to farming land has been argued by some sources as the sole motivator.[15] One opinion attempts to link Eureka to the racially exclusive White Australia policy.[16] Another portrays it as solely the work of foreign agitators and anarchists, the 'mongrel crew of German, Italian and negro rebels' vilified by the Sydney Morning Herald at the time.[17] Some historians described the events of Eureka as motivated by an essential demand that the miners be treated with respect.[18] Mark Twain compared Eureka to Concord and Lexington, conflicts that began the American War of Independence and eventually led to democracy in that land.[19]

Gravel Pits Ballarat 1854. The Gravel Pits diggings were near to the government camp. The use of the military to clear the riotous miners from the Gravel Pits at bayonet point during the ill conceived 'Digger Hunt' of 30 November was the spark that ignited the armed insurrection at Eureka. (Collection: Ballarat Fine Art Gallery)

Many of these interpretations leave much to be desired. While acknowledging the genuine passion of those who would claim that Eureka was the birthplace of Australian democracy, we can put aside their claims as little more than wishful thinking. If by democracy they meant the extension of the right to vote, democratic developments were occurring in other Australian colonies at the time. South Australia was a notable example, while Victoria itself had a constitution, awaiting approval in London, that granted an increased suffrage, albeit favouring the propertied classes. When considered in this context, Eureka cannot have been the birthplace of Australian democracy.

Yet this claim also misses the point of Eureka's true importance to the development of democratic traditions within Australia. While not the genesis of democracy, Eureka was its most robust and vigorous midwife. The events at Eureka climaxing in the bloody battle for the stockade set the criteria against which Australian democracy would be judged. After Eureka, any colonial government in Australia that ignored the aspirations of its ordinary people would know that it did so at its peril. It was in this way that Eureka forced those governments to back away, albeit reluctantly and with much dragging of feet, from the exclusivist, class-biased political reforms they had planned, and to adopt a much more inclusive approach. Eureka indelibly stamped an egalitarian mark on the character of the democracy that developed within Australia over succeeding decades.

Hunger for land and economic frustration certainly created an environment for dissent. They were not, however, issues unique to Ballarat, and had not caused an armed uprising anywhere else. There was no reason that those issues would, for some reason peculiar to Ballarat, lead to such an uprising. Interpretations of the causes of Eureka that are driven by rigid political ideology, such as those offered by Marxists and anarchists, are far too limited in their focus and not particularly useful.

Similarly, dismissing the Eureka miners as nothing more than greedy, self-centred and self-interested tax evaders is such an uninformed and empty-headed interpretation as to be unworthy of serious rebuttal. We can also discount the absurd claim of Eureka being in some way connected with the White Australia movement. The Eureka miners were certainly men of their times and no doubt harboured many notions about race and ethnicity that we would now consider unacceptable. However, there is no evidence that they were motivated in any way by such passions, while at least one of the insurgents was an African-American and another came from the West Indies. What then was the cause of the Eureka uprising?

Modern historians have put a great deal of thought into finding the answer to that question. Weston Bate stated that 'Eureka merely regained for the goldfields British civil liberties that three years of makeshift and often arbitrary rule had denied'.[20] John Molony proposed that the fight at Eureka was essentially a consequence of the denial to the miners of their right to be treated with respect.[21] In summarizing what he saw as the prime motivators for the uprising, Geoffrey Serle identified the 'fundamental irritant' as both a fight for freedom and a democratic protest against arbitrary government.[22]

There was among miners and those observing events on the goldfields a deeply held conviction that British subjects were entitled to expect their government not to resort to the use of armed coercion to enforce the law. For many people at the time such behaviour bespoke of despotism and would not be tolerated.

Those people who had flocked to Victoria in the quest for gold were characterised by the somewhat sanctimonious colonial establishment as being little more than a polyglot host of essentially vulgar treasure seekers. Why indeed should they not have accepted their lot and reconciled themselves to deserving no better than they got? Surely it would have been easier and more profitable for such people simply to comply with what was demanded of them.

Yet they did not comply. Instead they chose to resist, mobilised an armed force, and challenged the formidable powers arrayed against them. Determining what lay at the root of their decision to do so reveals much about the character of the men who defended the Eureka stockade, and helps explain why the battle developed into a serious clash of arms.

In the 1850s the terms *respectable* and *Englishman* were held by many to be synonymous. The comfortable, parochial, complacency and self-esteem enjoyed by a mid-nineteenth century British subject rested on several factors. The first was the knowledge that he was a free man, with certain rights assumed to be inalienable and protected by long standing tradition and convention. Ambiguous as such rights might often seem, in a culture of rigid social conformity they were heartfelt nonetheless.

The second feature of what made an Englishman resentful of being treated in an obnoxious manner was one peculiar to the diggings of colonial Australia. The hosts who flocked to the goldfields came from every strata of society. They were not normally destitute, and were generally literate, with many of them adequately educated. Such men, now in a land offering unbounded opportunity and half a world away from the constraints of their previous lives, would have heartily agreed with John Sherer, an English gold seeker who wrote that 'all aristocratic feelings and associations of the old country are at once annihilated ... It is not what you were, but what you are that is the criterion'.[23] In January 1852 the distressed property owner Alfred Burchett wrote that outsiders 'cannot imagine the state of things here. Men who have been servants all their lives are now, after a few weeks work at the diggings, independent'.[24]

The notion of individual independence was indeed important. Those who had come to Australia to seek gold had done so voluntarily. In the great majority of cases this meant a long journey across vast oceans, a significant expression of individual initiative as well as an expenditure of time and funds. On arrival in Australia, even if not themselves British, they could expect, as free immigrants and residents within a British community, that they were entitled to certain protections. Chief among these protections was that their independence would not be unfairly curtailed by arbitrary or capricious repression by those in authority. To do so was considered to be un-British, and a direct menace to the core values upon which their self respect and personal aspirations were based.

Unfortunately, during the events leading up to and following the battle for the Eureka Stockade, the perception among many miners and those observing events on the goldfields was that the colonial authorities were behaving in just such an un-British manner. It is important to realise that such sentiments were not an imaginative post-Eureka construction by apologists for the Eureka insurgents. They were not some selfish delusion invented by the insurgents themselves after the event to excuse rebellious behaviour. They were rooted within the very core of what were thought at the time to be British or English virtues. The denial of such virtues by the government of the day was the 'fundamental irritant' of Eureka.

Articulating just what were British liberties exercised many a brilliant mind and pen in the years before Eureka. In 1762 John Adams, the American lawyer and patriot wrote:

[l]et it be known that the British liberties are not the grants of princes of parliaments, but original rights, conditions and

original contracts, coequal with the prerogative, and coeval of government. That many of our rights are inherent and essential, agreed on as maxims and established as preliminaries even before parliament existed. Let them search for the foundation of British laws and government in the frame of human truth, liberty, justice, and benevolence, are its everlasting basis; and if these could be removed, the superstructure is overthrown of course.[25]

Even though penned 90 years before and half a world away from Eureka, Adams' contention that British liberties are inalienable, not conferred by princes or parliaments, and are everlasting ancient rights upon which the bedrock of British government is based, are directly relevant to the situation at Eureka. His warning that the very legitimacy of government is overthrown by the removal of those rights presents the prima face case for justifying insurrection. Such was the case at Eureka.

It is fortunate for us that the traditions and conventions related to the rights of an Englishmen, to which Adams referred and to which the Eureka insurgents responded, were considered to be an integral part of what made up English and subsequently British society. This was so much so that in 1689 the English parliament saw fit to formally codify them. In that year an Act, *Defining the Rights and Liberties of the Subject and Settling the Succession of the Crown of the English*, colloquially known as the English *Bill of Rights*, was promulgated. In this Act the English parliament set forth articles that strictly curtailed the impositions the monarch could impose upon an Englishman. In its lengthy preamble the Act referred to the, 'ancient rights and liberties' and the 'undoubted rights and liberties' of the English people. By doing so it confirmed that there was indeed a general perception, a social contract, among the English people that there were long established liberties inherent to all Englishmen and incumbent upon those who ruled them.

In a similar fashion the earlier Act of *Habeus Corpus*, enacted in 1679, established the right of an individual to be charged and tried in open court. This Act protected Englishmen, at least in theory, from arbitrary arrest and the clandestine exercise of judicial power. This affirmed the ancient right that an Englishman possessed a certain individual sovereignty, which could not be violated without transparent legal process.

While a few of the Eureka miners may have been able to quote aspects of the *Bill of Rights* and *Habeus Corpus*, presumably most would not have been able to do so. Nevertheless, all of them would have possessed an inherent understanding of the rights and liberties guaranteed to them by both Acts. Such notions were intrinsic to their understanding of it meant to be British. In consequence, any behaviour that abrogated these rights was seen as the difference between British and un-British behaviour. What then were these ancient and undoubted rights and liberties that the Eureka insurgents felt were being so cruelly abused?

The right to petition, which is in effect the right to be heard, was and remains a fundamental liberty enshrined in British law. By clearly stating that, 'all commitments and prosecutions for such petitioning, is illegal', the *Bill of Rights* acknowledged the right of the common people to petition the crown. This not only recognised that the common people should be heard, but that they could do so without redress or punishment from Monarchs or their representatives.

By granting this right, British law accepted that the opinions of the ordinary person had value, and by so doing affirmed the sovereignty of the individual. It also made it a legal requirement that a British government could not remain aloof from its subjects and rule without granting them at least the opportunity to put forward their point of view. This was a notion that was certainly not lost on the insurgents at Eureka or the newspapers of the times. Explaining the unrest on the goldfields in 1854, the Melbourne newspaper the *Argus* made the observation that the 'cardinal grievance of the diggers is their exclusion from any share in the representation of the country. Give them their fair share of this and they will be satisfied to trust to subordinate grievances being rectified by the legislature'.[26]

One example that shows that those on the Victorian diggings appreciated perfectly well their right to be heard under British law was the great Bendigo Petition of 1853. In that year the miners of the Bendigo goldfields, who had long suffered under the same coercive restrictions as those encountered at Ballarat, formalised their grievances in a written petition containing thousands of signatures to the then governor, Sir Joseph Latrobe. The response from Latrobe was underwhelming. He essentially ignored the petition, presaging by one year the refusal of his successor, Hotham, to countenance any negotiations with miners who were so bold as to affront the crown's representative with such words as *demand*.

To give Hotham his due, he did claim that he had attempted to conciliate the delegation by asking them to send their petition and let him see it. He was, however, by nature and profession a stiff backed military man, and his conception of the proper order of things balked at any more intimate communication with the discourteous miners.[27] This was particularly ironic as it was a specific liberty granted by the *Bill of Rights* that the people be allowed to 'demand of their rights'.[28] For a great many Victorian gold miners the dismissive attitudes of successive colonial regimes to their demands to be heard would have confirmed that, apart from them being a source of revenue, their governments had at best little interest in them, and at worst held them in contempt.

The *Bill of Rights* is equally clear on the limits of a government's coercive power. There was a limit set on the amount of bail and fines so that they were not excessive, and specific mention was made that punishments inflicted were not to be cruel or unusual. One punishment employed by the police in particular, that of chaining men to trees and logs in the open for failure to produce a valid miner's licence, was singled out repeatedly by aggrieved miners as both cruel and unusual.

For the miners, free men all, being chained like beasts to a log was seen as an extraordinarily obnoxious sanction, and it infuriated them. Peter Lalor, leader of the Eureka insurgents, wrote a year later that the 'diggers were subjected to the most unheard of insults and cruelties in the collection of this tax, being in many instances chained to logs if they could not produce their licence'.[29] John Bird, a miner at Ballarat in 1854, expressing his sense of outrage at the practice wrote that 'manly men of noble mien, in their shirts rolled up with bare and brawny arms, and with sashes round their wastes, marched like bushrangers between troopers with loaded carbines, bearing insult as only Britons could bear it'.[30]

Recalling the events of Eureka, John Stewart described the strength of feeling against the practice and lamented the lack of trust exhibited by such actions.[31] Pierson, an American miner who lived and worked on the Ballarat diggings at the time, thought that they were being hunted like game.[32] Carboni describing the 'digger hunts' that led to arrests which resulted in men being chained to logs appealed '[a]re diggers dogs or savages, that they are to be hunted on the diggings'.[33] Even Police Commissioner Charles MacMahon lamented that the 'system of securing to trees, men who had been arrested for non payment of the licence fee, was a disgrace to any civilized community'.[34] The Californian Ferguson echoed Carboni's cry when he wrote that the miners were treated, 'more like dogs than Christian Gentlemen'.[35]

The reference to 'gentlemen' is significant. Men who thought of themselves in such a manner were possessed of a certain degree of self-identity, dignity and self-esteem. To the despair of many in the entrenched Victorian colonial hierarchy, such men proved not to be deferential forelock-tugging menials, but forthright, articulate and independently minded. It was surely a dangerous thing to inflict demeaning punishments upon such men, especially those who considered themselves the inheritors of British rights.

The tensions created by the regime on the goldfields were apparent for all to see. In early October the *Argus*, warning of the dangerous passions being aroused among the miners, reported that 'there is a vast amount of silent resentment at the wrongs inflicted and systematic ill-treatment under the name of the law, smouldering in the bosoms of the gold diggers'.[36] In a summation some days after Eureka, the newspaper asked what 'else could have been expected from the exposure of freemen to the daily and hourly insult from a demoralised, but almost absolute, police force?'[37]

To expect protection from corruption and incompetence among public officials was another right guaranteed by British law. Corrupt and unqualified persons were not to be placed in positions of legal authority. Both failings abounded on the Ballarat diggings where the behaviour of a few notoriously corrupt individuals set the standard against which all were judged.

In an environment in which the police were compelled to enforce a series of laws that had almost no public support on the goldfields, it was inevitable that the police would make few friends. Some of the men recruited into the police were not up to the standards required. This was inevitable in an environment where replacing the many serving police who had resigned to go and seek their fortunes was a priority for the authorities. In such circumstances it would not have done to enquire too closely into the backgrounds of those who were willing to answer the call.

Even for those police whose backgrounds were unsullied, there was a constant struggle against the temptation to play favourites in an environment that demanded the enforcement of obnoxious laws. One official report had described the result of the police administering the alcohol and miner's licence laws as having 'a most pernicious effect on the morals of the police force'.[38] Turning a blind eye at the right moment could often prove to be financially rewarding, and it was not unknown for police to supplement their meagre salaries in this manner. The conduct and outcome of Bentley's murder trial simply confirmed for a great many at Ballarat the depths of dishonesty to which their public institutions had sunk. Lalor, who had helped organise a committee to deal with the Bentley issue, cited the outcome of this trial as a specific reason for the uprising, and he was correct.[39] For Lalor and all the other aggrieved men and women on the Ballarat fields, their intrinsic rights had been abrogated.

While official corruption inflamed passions, the perceived incompetence of many officials also caused serious disquiet. The poor administration of the goldfields and the colony in general had been a cause for unrest even before Eureka. Petty officials on the goldfields had a fundamental role to play in the daily lives of the miners. It was imperative that such men knew their jobs and were able to administer them in an efficient and competent manner.

Unfortunately, far too many were seen as 'young ignorant fellows, who are much too delicate and fine to do anything', while others were criticised as inexperienced young fellows who had been put into positions of authority when far more competent men had been passed over.[40] Despite such observations, there were competent officials working on the goldfields at the time. Gold Commissioner Gilbert Amos, who was intimately involved with the Eureka uprising on the government side, was one such capable and skilled official. However, as with all such things, it is the general perception of incompetence that mattered. A letter written to the Boston *Daily Atlas* by American miner John Fisher in 1856, lambasted Latrobe as 'an ignorant, stupid, one-ideaed [sic] lump of arrogance'.[41] Pierson described those whom the miners' taxes supported as 'a lot of miserable lucky vagabonds in office'.[42] Prominent insurgent Frederick Vern, reviewing the reasons for the uprising, wrote in November 1855 that the miners had been slaves to 'official insolence and petty authority'.[43]

In normal circumstances such deficiencies might be remedied by complaints, rebukes and replacement, but in the Ballarat of late 1854 this was not possible. Two successive colonial governments had refused to listen to the miners' grievances when presented in a manner that complied with the expectations of public remonstrance. The petitioners felt they had a right to be heard, and it had been denied to them. A further match had been set to the fire upon which simmered the pot of violent upheaval.

There was one liberty confirmed by the *Bill of Rights* that would have direct relevance to the Eureka uprising: that of bearing arms in one's own defence. Couched as it was in the language of the later seventeenth century, and justifying the law as a means for Protestant England to defend itself against the feared machinations of Roman Catholicism, the *Bill* nevertheless unambiguously conferred the right of individuals to bear arms to protect themselves from those who would do them harm.

Legal opinion had confirmed that the right to defend oneself was an extension of English Common Law, the ancient law of the land. In his charge to the Bristol Grand Jury following the riots in that city in 1832, Justice C.J. Tindal sanctioned the right of self defence, stating that when 'violence occurs, every subject has the common law right to use force against force'.[44] While further opinion given by Tindal made it clear that this right did not extend to those engaged in resisting the lawful authority of the state, the concept of self defence was something the miners would have understood as an inherent right.[45] The parading of soldiers with loaded muskets and fixed bayonets when Wise had provocatively marched his armed company into Ballarat on 28 November certainly could have been seen by the miners as a threat to use force. The clash at the Gravel Pits on 30 November had confirmed those fears.

In a letter to Alicia Dunn written on 30 November, Lalor stated unambiguously that self-defence following the firing on the miners by soldiers at the Gravel Pits was the reason for taking up arms.[46] Military displays were guaranteed to cause great dismay among those who considered they lived under the protection of British law. Many incidents attest to this. Shortly after the fall of the stockade, the correspondent at Ballarat for the *Argus* was passing near the government camp when he was confronted by a trooper who, sword drawn, asked the correspondent and his friend if they had any right to pass that way. The policeman asked politely and made no difficulty once the correspondent and his friend had given an account of themselves. Yet the correspondent was uncomfortable enough with what had happened to observe that this was something 'to which Englishmen are not accustomed'.[47]

Capturing the mood of the diggings at the time, the *Argus* wrote that the 'appearance of the military at Ballarat will doubtless be offensive'.[48] Indeed it was, causing Reform League leaders John Basson Humffray and Thomas Kennedy to warn that if the army or police continued to threaten the people without reading the Riot Act, then the people could be expected to defend themselves.[49] That Humffray, a consistent advocate of non-violence and a constitutional solution, was moved to threaten to meet violence with violence illustrates the gravity with which threats from the military were viewed. In commenting on the Eureka uprising, the *Australian and New Zealand Gazette* observed that the actions of the government at Ballarat were challenging the people to fight, which as Englishmen they would do regardless of the fearful odds against them.[50]

For a British government to use military force to suppress civil disorder was not unusual. There were numerous examples of British authorities using soldiers to deal with dissent and unrest in Britain and in other colonies to provide a precedent for what happened at Ballarat. Before the creation of an effective civil police force in Britain, the army was the ultimate means of maintaining law and order, especially outside the large cities. The use of the army to restore order forcibly during civil and political disturbances throughout Britain during the eighteenth and nineteenth emphasised the frequently ambivalent relationship that existed between the army and the vast mass of ordinary people in Britain during the early nineteenth century.[51]

Outside Britain, the army had been used to repress bloodily a rebellion in Canada during 1837, and it regularly did the same throughout Britain's vast and still growing Indian possessions. When soldiers arrived to garrison the government camp at Ballarat, and were then reinforced, astute observers on the diggings would have been aware of the implications.

It was the threat of military violence in the form of the bayonet, the iconic weapon of despotic coercion, which seemed to aggravate passions the most intensely. Carboni, describing the sense of outrage at the threatened use of that weapon, wrote that 'John Bull ... was born for law, order and safe money making on land and sea ... he hates the bayonet: I mean of course that he does not want to be bullied by the bayonet'.[52] Lalor called the threat of the bayonet 'an insult to [our] manhood'.[53] The *Australian and New Zealand Gazette* wrote of what it deemed to be the undoubted British right to resist demands on a man when made 'with a fixed bayonet at his breast'.[54]

When Humffray and C.F. Nicholls presented a petition to Hotham in January 1855, they made special mention of the weapon, stating that they wished to 'condemn in the most unqualified manner the conduct of the Ballaarat officials in collecting a tax at the bayonet's point'.[55] In a reference to the means of tax collection used by the government against the miners, the *Ballarat Times*, a radical supporter of the miners' cause, wrote on 28 October 1854 that it 'is not fines, imprisonment, taxation and bayonets that is required to keep a people tranquil and content'.[56]

For Carboni, himself a refugee from the political violence and revolution of his native Italy, this was uncomfortably reminiscent of 'the hated Austrian rule, which was now attempted, in defiance of God and man, to be transplanted into this colony'.[57] In a further rhetorical flourish written one year after Eureka, Carboni re-emphasised the alien imposition of military coercion upon a British population, exclaiming that 'I came then 16,000 miles in vain to get away from the rule of the sword'.[58] Such was the sense of dread that pervaded on the evening of 30 November when the miners of Ballarat gathered together under the banner of the Southern Cross on Bakery Hill.

It must be always remembered that even at this stage there was never any consensus reached among the aggrieved miners on how they should best proceed in the face of military and police armed intimidation. The Reform League was split into two mutually antagonistic factions, that of *Moral Force* arguing for non violent constitutional change, and *Physical Force*, whose philosophy was best illustrated by the admonition of one of its leaders that 'moral force is all humbug, nothing convinces like a lick i' the lug'.[59] Opinions remained polarised until the 'digger hunt' of 30 November, when, faced with the imminent prospect of the use of military force against them, many sided with the Physical Forcers. The Moral Forcers attempted to reconcile passions, but to no avail. Having lost their audience for the moment, they slipped quietly into the background. An even larger number of others stepped back to the sidelines, to watch and wait for whatever might eventuate.

It was Butler Cole Aspinall, defence counsel for the African-American John Joseph at the State Trials held after Eureka, who most eloquently articulated the fundamental cause of the uprising at Eureka. In addressing the reasons his client took up arms he stated 'that though a negro, in any British possession he was entitled to his liberties'.[60] In just a few words and using his client as a living example, Aspinall illustrated in a simple yet incisive manner that there was indeed among British subjects an expectation of entitlement to respectful treatment from their governments. Aspinall made his point explicitly, stating that even the humblest subject, a negro, one of a people who in the unenlightened racial context of the times were considered to be deficient in regard to intellect and morals, was nonetheless in possession of inalienable liberties guaranteed by British law. Aspinall's point is clear; if such a person could expect to be treated with respect by those in power, then so indeed should everyone.

Swearing Allegiance to the Southern Cross 1854. In this contemporary watercolour Charles Doudiet, who was on the Ballarat diggings at the time, recorded the historic moment when several hundred armed miners gathered beneath their Southern Cross flag and swore to defend each other and their rights and liberties. (Collection: Ballarat Fine Art Gallery)

It was in this context that the Eureka miners began reaching for their pistols, revolvers, shotguns and rifles. For those without such means the word went out to 'provide themselves with a piece of steel … that will pierce the tyrant's heart'.[61] When hundreds of men gathered in armed array beneath the Southern Cross flag on Bakery Hill, they knelt as one and removed their hats. Lifting their faces to their starry banner they swore 'to stand by each other and fight to defend our rights and liberties'.[62] For the men who took the oath, this was no casual boast, idle whimsy or pretentious bombast. It was simply stating what had brought them to that place and the nature of the conflict to come.

Chapter 2

'GENTLEMEN SOLDIERS': THE INSURGENT ARMY

Aroused to insurrection, the gold miners formed into divisions, which in the military terminology of the day meant companies. These gentlemen soldiers, as they called themselves, segregated their companies according to the weapons they brought with them. Some carried pikes, some double-barrelled shotguns and fowling pieces, some pistols and revolvers, while others carried rifles. Companies were formed composed of men from all the nations represented on the diggings, at least one being made up of mostly Italians and French.[1] Another was made up of sailors whom it was thought best to keep them together because they fought in their own peculiar style. A German band played martial music. Captains were appointed, one for every company.

Carboni was there, armed with a sword at the head of the Italian and French insurgents. The German Vern, trailing a large sword, was prominent. The Canadian Charles Ross took charge of a division of men armed with rifles and muskets. Irishmen Michael Hanrahan and Patrick Curtain commanded a division armed with pikes, the iconic weapon of Irish rebellion.[2] Edward Thonen, a Prussian who had lived through the revolutions in Germany during 1848, took charge of another company of riflemen. An American carpenter named Nelson, or Nealson in some accounts, commanded a division known as the First Rifles by one account and the Californian Rifles by another, a group of men described as 'the very best'.[3]

Up to 1500 armed men mustered and began drilling on 1 and 2 December, the Friday and Saturday before Eureka. To an unpractised eye it no doubt looked to be a formidable array, certainly not one to be dismissed lightly. Unfortunately for the insurgents their military pretensions were little more than a hollow illusion, the consequences of which would be catastrophic defeat. Yet it would be a mistake to belittle or underestimate everything of their military preparations or dismiss out of hand their martial performance on the day of the battle.

Much has been made over the years of the painfully inadequate military situation the insurgents found themselves in at Eureka. Great emphasis has been placed on how the very poorly armed gold miners were set upon by a fully armed and brutally repressive regime. As expected, this is the interpretation of events that is favoured by those partisan to the miners' cause, and it has captured the romantic imagination. It is an interpretation that bears little resemblance to the circumstances. What then was the actual state of affairs?

Lack of firearms and limited availability of ammunition to the defenders of the Eureka Stockade, or that most were armed only with pikes have been articles of faith in the Eureka story.[4] An examination of the actual evidence, however, reveals that the insurgents inside the stockade were much better armed than folklore insists.

The number and type of firearms available to the Eureka insurgents would have been governed by what was to hand. When trying to assess what firearms might have been available, the observations of those who lived and worked on the goldfields during the early 1850s are illuminating. James Arnot, describing the journey of his party to the Bendigo diggings in 1852, related how they were armed:

for our protection I carry a brace of pistols, camp knife and preservers, Peterson [carries] a rifle and Bowie knife, Bettison a brace of pistols and Kelly a ball gun, that is the armour of my private party but among the group consisting of 30 or 40 there are plenty of rifles.[5]

Peter Lalor. The commander in chief of the insurgents, Peter Lalor acted as a unifying influence for the multinational insurgent 'army'. During the battle Lalor bravely stood his ground until severely wounded following which he was hidden from view and managed to escape after the battle. Lalor later reconciled with the political establishment and went on to become Speaker of the Victorian parliament. (State Library of Victoria)

The Ballarat correspondent for the *Argus* wrote that at the meeting of miners held on 29 November 1854, some 'thousands of guns and revolvers were fired off at one period'.[6] Carboni confirmed this when he wrote that during the same meeting a 'regular volley of revolvers and other pistols now took place'[7] Writing in 1853, Mrs Charles Clancy mentions a typical night on the diggings, recalling 'revolvers cracking - blunderbuss bombing - rifles going off - balls whistling'.[8] The *Mount Alexander Mail*, reporting from the Ballarat diggings at 6 a.m. one morning a few days before Eureka, described how 'the report of firearms is to be heard in all directions'.[9] Lazarus mentions a fearful racket of many guns being fired off by the miners through the night of 30 November.[10]

Samuel Huyghue, a government employee at Eureka, when describing the night of 28 November when the soldiers of the 12th Regiment were attacked, mentions the incessant flashes from guns and revolvers being fired into the sky outside miners' tents, lighting up the darkness.[11] William Howitt, writing from the McIvor diggings in 1855, states that 'guns … are an especial nuisance. They are continually discharging them on the diggings'.[12] These accounts all indicate most strongly that a great many firearms were in the hands of the gold mining population. The *Argus* and *Mount Alexander Mail* reports, and those of Lazarus and Huyghue, are especially important as they specifically refer to the Ballarat diggings.

Despite what appears to be evidence of the widespread availability of firearms, some among the Eureka insurgents made special mention of their supposed shortage. Lalor gave an account of the number of firearms in the possession of the insurgents on the morning of the battle, stating that there 'were about seventy men possessing guns, twenty with pikes, and thirty with pistols'.[13] Carboni made the point that in the days leading up to the battle, '[a]rms and ammunition were our want'.[14] John Lynch, who was inside the stockade, echoed the sentiment, recalling that there was a 'scanty supply' of firearms and ammunition. As will become apparent, all of these claims are exaggerations. Lynch, however, does mention the 'disparity of arms' between the attackers and the defenders.[15] He was most likely referring to the quality of weapons available, rather than the numbers in the hands of the insurgents. This was a much more accurate means of assessing the relative armaments available at Eureka.

There has also been a persistent claim in many sources that the insurgents suffered a shortage of ammunition. Lalor stated that those defenders possessing firearms each had 'no more than one or two rounds of ammunition'.[16] The oft-repeated account by W.H. Fitchett of a pistol loaded with quartz pellets being found inside the stockade has also been used as further proof of the insurgents' lack of ammunition.[17] Such claims are, however, quite misleading. Lalor could be expected to down play the military prowess of the men he led after the event, while Fitchett, who was not at Eureka, was merely repeating hearsay. The most obvious evidence that the Eureka defenders were adequately armed is in the sequence of events that occurred after the fighting began. Fire that was kept up for roughly ten minutes, and proved capable of causing significant discomfort to the attacking soldiers, could not have been produced by men who lacked firearms or ammunition.

Armed miners on rout (sic) to deposit gold 1852 – 53. The personal firearms carried here show clearly that such weapons were not at all uncommon amongst the miners. Many accounts describe miners carrying and using firearms ranging from pocket pistols to revolvers and shotguns on a day-to-day basis. (The University of Melbourne Art Collection. Gift of the Russell and Mab Grimwade Bequest 1973.)

When assessing what arms were available to the Eureka insurgents, it is important that we define what is meant by the term 'arms'. In the context of Eureka, 'arms' meant firearms of a specific type. These were muzzle-loading, single shot, long barrelled and of sufficient calibre to provide an effective counter to the muskets of the military. When we read accounts of the insurgents lacking arms, this is the type of firearm being discussed. Military muskets and bayonets, the weapons the soldiers carried, were not normally included among the accoutrements of a typical gold miner, and for very understandable reasons the insurgents would have felt themselves deficient in this regard. There were, however, a great many weapons available to them. Collecting enough of the right types to make a difference was their challenge, and when one considers the determined exchange of fire that occurred for a significant period of time on the morning of the battle, it does seem that they had succeeded at least to some degree in meeting that challenge.

The types of firearms that would have been accessible to the Eureka insurgents were certainly diverse, so much so that there was no possibility of them being armed in anything like a uniform manner. A great many miners on the goldfields carried a handgun of some type. They were common enough to be referred to by one correspondent as 'a handy weapon'.[18] In an environment where law enforcement was at best intermittent, and lawless behaviour was not an infrequent occurrence, it would have been the foolish miner who did not have at least shared access to a handgun.

Possession of such weapons also provided those gold miners so inclined with a more concrete guarantee of their personal independence and protection from over-zealous officialdom. The lyrics of *The Maryborough Miner*, a ditty from the era, emphasised the fear of the police for the recalcitrant miner's revolver, which he referred to as his patent pill machine.[19] Many of these handguns were pistols, often single shot and of small calibre, with limited stopping power, and very inaccurate over more than a few yards. Revolvers, as alluded to by *The Maryborough Miner*, were also available and highly prized.

Several types of revolvers were available to miners on the diggings in 1854. The most common was the six-barrelled Pepperbox style, but without doubt American Colt revolvers were the most popular. When ships from the United States dropped anchor at Melbourne, it was not uncommon for people to board them and ask to buy the passengers' Colts.[20] They were reliable and robust. One drawback, however, was that with the exception of the .44 calibre Walker Colt, the man-stopping power of a single shot was unpredictable. Unless lucky, or in the hands of a well practiced, deadly accurate shot, one or even two balls fired by the lighter calibre Colts would normally not be enough to ensure a kill, or even to incapacitate an enemy. There is a report of a British officer putting six balls from his .36 calibre Navy Colt into the chest of an attacking mutineer during the Great Sepoy Rebellion of 1857. The six shots did not stop the attacker before he was able to strike the officer with his sword and kill him.[21]

Armed miners working their claim. Even when working their claims, miners were sure to never be too far from some means of protecting themselves and their property. This miner has a pepperbox revolver in his belt, a not at all uncommon occurrence. (Ferguson, The Experiences of a Forty Niner during a Third of a Century in the Gold Fields.)

From descriptions of those who saw them, most of the Californians at Eureka were armed with Colt revolvers, and their presence inside the stockade added significantly to the armament of the insurgents. One, Ferguson, admitted to carrying a Colt, which he slid down his trouser leg onto the ground and kicked aside before he was searched following his capture.[22]

In the popular imagination of the time, and in reality, Californian and American were terms synonymous with revolver.[23] In 1855 the *London Illustrated News*, deploring the number of homicides that had occurred on the Californian diggings that year, attributed them to 'the universal and cowardly practice of carrying revolvers'.[24] Lord Robert Cecil, a future British Prime Minister, described an American with whom he shared a coach ride while visiting Victoria as having 'a pair of pistols in his belt'.[25] Carboni mentions the Californians arriving at the stockade, each one 'looking Californian enough, armed with a Colt's revolver of large size'.[26]

William Kelly, describing those Americans he met on the Ballarat diggings in 1854, mentioned them being 'girthed above the hips by a red sash, that was stuck round with knives, daggers, and revolvers'.[27] When asked only a few years after Eureka if the day of the revolver had passed, one Californian on an Australian gold field answered '[n] ot with us ... these little instruments have wrung from the oppressors that justice which would have been withheld'.[28] Even if many of these descriptions were from people who may simply have been relating their preconceived impressions of Americans, the consistency of the descriptions does leave a definite impression that there was a definite connection between them and their revolvers.

The actual prevalence of revolvers of any type among the miners in general, and the Eureka defenders in particular, is difficult to determine. Burchett, writing about life on the diggings, claimed that 'everybody at least carries a six barrelled revolver with him'.[29] This may have been an exaggeration, but it does indicate that such weapons were common enough for a casual observer like Burchett to notice. William Gay's father certainly possessed two Pepperbox revolvers. Reports of criminal behaviour on the goldfields frequently mention revolvers. Pierson recounts how bushrangers robbed diggers by threatening them with revolvers. Just before Eureka, he noted in his diary that he could see hundreds of miners assembled on Bakery Hill, some of whom were armed with revolvers. [30] A Mrs Thomas, pregnant at the time of the Eureka troubles, was in her tent and armed with a revolver with which it was said she shot a thief.[31] One Castlemaine miner recalled how a man had stepped into their tent one night, produced a revolver, and attempted to rob them.[32]

The Reverend Theophilus Taylor recorded in his Ballarat diary for 2 December 1854 that he witnessed what he called rebels patrolling the diggings armed with revolvers and other weapons.[33] Mrs Clancy's mention of revolvers cracking each night on the diggings is worth recalling. Insurgent miner Lynch claimed to be armed with a revolver, but does not mention its type.[34] One Eureka defender, referred to only as 'One of Lalor's Captains', mentioned firing his revolver at two soldiers, but does not mention the weapon's type.[35]

It remains uncertain, though, if revolvers were as common as simpler types of handguns among the general digger population. Lalor carried a double-barrelled muzzle-loading percussion cap pistol.[36] That the leader of the insurgency was so armed may have meant that he was unable to procure a revolver, or perhaps he was just not interested in having one. It does seem, though, that revolvers were common enough on the diggings to make a significant impression upon those who lived and worked there. Accounts by those at the stockade and the presence of the Californians there indicate that revolvers would have been available in significant numbers.

Despite their versatility at close quarters, revolvers were of little use in any serious confrontation with soldiers armed with muskets. Longer ranged, larger calibre weapons were needed to redress the balance. Luckily for the insurgents such weapons were not uncommon on the goldfields.

The Ballarat correspondent for the *Argus*, reporting on a meeting of about 2000 armed miners, wrote that they carried all sorts of weapons among which he saw 'the rifle of Manton … the cheap Birmingham fowling piece … the djerid of the Arab',[37] a true menagerie of firearms.

Many had access to 'double barrelled guns' and 'fowling pieces', the terms 'gun' and 'fowling piece' being the vernacular of the era for shotguns of various calibres.

There were numerous eyewitness references to the use of shotguns by the insurgents. Constable Hugh King of the foot police found shotguns lying about inside the stockade following its fall.[38] The African-American insurgent Joseph was seen by numerous witnesses carrying a double-barrelled shotgun.[39] Lynch and Michael Tuohy each carried a double-barrelled shotgun.[40] Lieutenant Thomas Richards and Private James Lough of the 40th Regiment mention seeing men armed with double-barrelled shotguns inside the stockade as the attack began.[41] Inspector of Police Henry Foster and police spy Henry Goodenough recalled seeing insurgents drilling and carrying their shotguns in the days before the battle.[42] Lalor did claim that 70 of the defenders of the stockade possessed them, and threatened to use his on any of the insurgents found looting.[43] It is evident that there were plenty of shotguns available to the stockade's defenders. What then of other long arms?

Muskets and rifles were also available. Muskets were heavy barrelled smooth bore weapons capable of firing a substantial solid bullet. Rifles had grooves machined into the bores of their barrels, to impart a spin to the bullet, ensuring greater accuracy over longer ranges. Carbines, a shorter version of the musket, were also found inside the stockade following its fall.

Several sources confirm that the insurgents had access to muskets and rifles. Carboni mentions Lalor holding his rifle with his left hand and resting its butt on his foot when he administered the Southern Cross oath to the assembled armed miners on Thursday 30 November.[44] Two military muskets were taken from the carts of the 12th Regiment when diggers attacked them on the night of 28 November,[45] and Constable Joseph Glover, who was stationed at the Eureka government camp, stated that a party of insurgents took four muskets from the camp.[46] Robert Burnette, an American digger, was armed with a rifle.[47] Private Patrick O'Keefe saw Joseph pick up a rifle after dropping his shotgun.[48] Trooper and police spy Andrew Peters testified that he had seen John

Manning, the correspondent for the radical *Ballarat Star* newspaper, drilling with many other men armed with rifles on 30 November and 1 December.[49] The formation of Nelson's First Rifles implies that there were enough rifles available for at least one distinct unit to be described as being armed with them. Sergeant Major of Police Michael Lawler and police Trooper John White reported seeing miner Henry Read armed with what they called a 'long piece'.[50] Neither Lawler nor White indicated if this long piece was a shotgun, musket or rifle.

The effectiveness of these firearms varied. Shotguns fired pellets and shot of various gauges. These could range from light birdshot to more substantial buckshot, or even a solid lead bullet. Although the hitting power of a shotgun would be marginal at ranges beyond 50 yards (45 metres), its effect at very close range could be devastating. It was certainly not a 'toy', as suggested in one account.[51] Muskets, which are discussed in some detail in later chapters and Appendix 1, fired a single heavy ball. As smooth bore weapons they were notoriously inaccurate, but did possess great stopping power if they hit a target in a vital region. Rifles fired smaller bullets than muskets, took longer to load, but were far more accurate. In the hands of a good marksman they could be deadly at ranges well beyond those of the musket or shotgun.

The type of ammunition loaded into shotguns could improve their effectiveness. Sometimes both buckshot and ball, known as 'buck and ball', were loaded together in a musket or shotgun. This type of load could be particularly effective in extending the killing range of a shotgun. Lalor was hit in the arm by a musket ball and two smaller balls at the one time suggesting that some of the police or army may have been using mixed loads in their weapons. There is no evidence that any of the Eureka defenders used such ammunition, although some, especially the Mexican-American War veterans among them, would surely have known of it.

Despite the number of handguns and shotguns available to them, there was a pressing need for the Eureka insurgents to obtain arms and ammunition capable of matching those carried by the soldiers. To procure these they resorted to requisitioning. Pierson, who was working his claims at the Gravel Pits and had a tent in sight of Bakery Hill, wrote that on Friday 1 December, the insurgents were 'very busy collecting arms and ammunition etc.', adding that 'they pressed these from the stores or any one that had them'.[52] The numbers and quality of weapons collected were no doubt as varied as the sources from which they came.

Receipts issued by insurgents for goods seized give an idea of the items taken. One read 'Received from the Ballaarat Store 1 Pistol, for the Comtee X, Hugh McCarty - Hurra for the people'.[53] Another read, in quaintly semi literate style, the 'Reform Lege Comete - 4 Drenks, fouer chillings; 4 Pies for the fower of the neight watch patriots - X.P'.[54]. A more erudite example from James Esmond to the firm of Bradshaw and Salmon in Main Road Ballarat read 'Got from Bradshaw and Salmon on 30 Nov: 12 lbs Powder, 1 Pistol flask, 1 Box Revolver Caps promised £3.12.0 to pay'.[55]

Some foragers did threaten violence against storekeepers despite Lalor's stern admonition that he would shoot any thieves. No doubt many less than honest characters took advantage of the tumultuous situation. An employee of Bradshaw and Salmon claimed that an insurgent named Moran threatened to shoot him if he didn't, 'hand over quick'.[56] Amos reported that insurgent miners cocked and uncocked their weapons in his face when they apprehended him and looted his post near the Eureka diggings.[57] Carboni left a vivid description of an incident at the Prince Albert Hotel where a group of ten heavily armed men claiming to represent the insurgents entered and demanded free drinks. The men got their drinks but were then frightened off by the proprietor Carl Weisenhavern brandishing pistols and his partner Johan Brandt wielding a double-barrelled shotgun.[58]

Carboni also mentions one incident when a group of diggers robbed the store of D. O'Conner of Yorkshire ham and coffee as well as £20 from the store's cash box.[59] Thomas Allen, who had a store adjacent to the stockade and was not sympathetic to the insurgents' cause, reported that he was robbed of a shotgun.[60] The insurgents not only requestioned guns, powder, food and drink. Saddles and other horse furniture were also taken. W.H. Cooper, a store owner, claimed that after a forward party of foragers had been rebuffed, 20 armed men returned to take what they wanted, which included twelve saddles and accoutrements for horses.[61] Huyghue mentioned several prisoners being brought in to the government camp after they were apprehended requestioning saddles.[62] Half a dozen saddles were also taken from the auction room of Dan Sweeny.[63] Carboni recounted how two men riding near the stockade had their horses taken from them.[64] Just what the insurgents intended to do with these items was never explained. Such behaviour, even if committed by rogue elements, did little to further their cause among some of Ballarat's storekeepers.[65]

Unfortunately no account has survived that lists exactly what all the Eureka foraging parties collected, or how many firearms were included in their haul. However, the numbers of firearms and quantities of ammunition found inside the stockade after the battle does seem to indicate that the foragers were reasonably successful in their quest.

Determining the proportion of insurgents with firearms inside the stockade may not be as difficult as might be imagined. There are several accounts left by those who stormed the stockade that assist us to do this. Giving evidence at the State Trials, Sub-Inspector Charles Jeffries Carter of the Foot Police testified that when he entered the large tent inside the stockade known as the Guard Tent, he found that it contained 16 or 17 stands of arms.[66] A stand in this case referring to a weapon and its accoutrements, which in military terms consisted of ammunition box or pouch, bayonet, scabbard, sling, and belts for carrying the ammunition box and bayonet. In the context of the stands of weapons in the Guard Tent, a stand would probably have been the weapon and perhaps a powder flask. Richards confirmed that arms and ammunition were found inside the Guard Tent, as did Amos, who referred to ammunition stored in the burning Guard Tent exploding after the stockade had been stormed.[67]

As will become apparent later in this narrative, the intentions of many of the gold miners were not at all pacific. It is hard to imagine that in the circumstances they found themselves in, they would not have armed themselves with whatever firearms were available, implying that the weapons and ammunition found inside the Guard Tent were surplus stores.

One argument that is often put forward as proof the stockade's defenders were hopelessly underarmed is the use they made of pikes. Such an assessment is understandable but mistaken, as it fails to appreciate the usefulness of the pike as a weapon in the situation of the Eureka insurgents. Despite the comment by James McDowell, a timber merchant, that he was robbed of his horse by men armed with guns and old rusty bayonets, the vast majority of the insurgent diggers' long-arms could not be fitted with bayonets.[68] Nor did it seem that they had access to any significant number of bayonets.

This placed them at a serious tactical disadvantage. In the era of single shot muzzle-loading weapons, the bayonet was the weapon *par excellence* for close quarters fighting. The bayonet was also the preferred weapon for repelling mounted attackers, the one element of the government forces that the shotgun and rifle armed insurgents had no protection against when in the open. Once a muzzle loader not equipped with a bayonet had been fired, and the enemy closed in for the kill, the weapon reverted to being little more than a club. Against a fast moving man wielding a sword from horseback, or a foot soldier armed with a bayonet at the end of a sturdy musket, such a club would be of little use. Pikes helped to redress this.

The pikes available to the insurgents were made on site by numerous blacksmiths. Each was a rough wooden pole about eight feet (2.4 metres) long, tipped with a sharpened metal spike and rough hook shaped blade. The spike was intended for thrusting, while the hook was designed to cut the bridles and reins of mounted men. The hook could also be used to pull a mounted man from his horse, so he could be dispatched by some other means. Such weapons, although crude, were sturdy and certainly not to be despised, especially if used in a coordinated manner.

Allen, observing a group training with pikes, recalled the instructions being given to them. Commands such as, 'Shoulder arms … Order poles … Ground arms … Stand at ease … Pick up poles … Shoulder arms … Right face … Quick march … Right counter march'[69] indicated at least an attempt to instil some sort of coordination among those armed with pikes. The pikemen Allen saw drilled for about two hours, after which they were formed up into ranks three deep and were addressed by their 'Captain' Hanrahan, who told them how to repel cavalry, and advised them to '[p]oke your pike into the guts of the horse and draw it out from under their tail'.[70]

Allen was a veteran of Waterloo, and quite deaf. He may have been recalling what he thought he had heard, and putting words remembered from his own military past into the mouth of the pike drill instructor. But even if this was the case, he did see insurgents drilling with pikes, and that alone indicates that there was some sort of concerted effort to regulate the pikemen.

Another use for the pike would be in close combat with bayonet-armed soldiers or police. Montague Miller, a pikeman at Eureka, recalled that his pike easily outreached the soldiers' bayonets.[71] Pikes could also reach mounted men, and be used to kill and injure horses, as seems to have been their intent from what Allen claimed to have overheard from the insurgent pike drill instructor. Those who were armed only with muzzle-loading shotguns, muskets, and rifles without bayonets could not hope to achieve any of these things.

Even though it provided some counter against a mounted enemy, the pike was not an ideal weapon. A pikeman had no firepower at all, unless carrying a handgun. Even then, using their handguns would mean that pikemen would not be able to wield the pike effectively. Unless supported by insurgents with firearms, the pikemen would not be able to resist firearm-equipped infantry for very long, as was to be the case at Eureka. In the circumstances the Ballarat insurgent miners found themselves in, however, the pike was an effective compromise that addressed an immediate crucial need in a force that lacked close combat weaponry to match that of their enemies.

Other evidence also suggests that the defenders of the stockade were not as badly armed as popular myth suggests. When giving evidence at the State Trials, Amos recalled seeing the 100 insurgents who captured him immediately protect the firelocks of their arms when it began to rain.[72] Such actions would not have been necessary if the firearms had been handguns, which could simply have been put into a pocket, but would have been required if the weapons were long arms. Thus, there were at least 100 insurgents armed with shotguns, muskets or rifles. Of course not all, perhaps not any, of these men may have been inside the stockade on the morning of the attack. It may, however, be significant that the insurgent in charge of those men was Ross, whose division did form part of the garrison the morning the stockade was attacked.[73] Whether it was Ross's men Amos saw, or Ross just happened to be with them, we will never know.

It is apparent that there were firearms of many types present within the Eureka stockade. The double-barrelled shotgun featured prominently, as did handguns, including numerous revolvers. Some muskets may have been present, while there certainly were at least a few rifles. A significant amount of ammunition was stored in the stockade, as well as a not insignificant number of unused firearms. In addition to firearms there were pikes, which, as we have seen, were a sensible substitute for the bayonet in the circumstances.

One other very important indicator that there were adequate numbers of firearms and stores of ammunition inside the stockade was the weight of firepower generated by its defenders. That this fire was kept up for about ten minutes and caused noteworthy discomfort to the attacking Redcoats is significant. Such an outcome would not have been possible if there had have been a chronic lack of firearms and ammunition inside the stockade.

Forming effective armies from masses of civilian volunteers has always been a challenge, particularly when the time available is limited and those attempting to create such armies are amateurs themselves. Such was the case for the Eureka insurgents. Civilians are by their nature unused to the subjugation of self-interest demanded by the military. Keegan reinforced this point when he wrote that militarism 'deprived free men of their right to protest, to demonstrate, to heckle, to jostle, to intimidate, to riot'.[74]

Such a description fits perfectly the free spirited and rambunctious nature of the defenders of the Eureka Stockade. The process of transforming such independently minded individuals into soldiers is neither simple nor quick. Yet it was essential that it be done, especially in a military era in which the weapons used demanded coordinated employment to be effective. Without this, there could be very little chance of achieving success.

It is a sad irony that it was the very individualistic spirit among the miners that provided the kindling for the flame of rebellion among them that constituted their greatest military weakness. When they did decide to act, and gathered together, their muster resembled little more than a crowd, which is exactly what it was. Keegan neatly captured the essential spirit and character of such crowd-armies when he observed that inside:

> every army there is a crowd struggling to get out ... the crowd is the antithesis of an army, a human assembly animated not by discipline but by mood, by the play of inconstant and potentially infectious emotion which, if it spreads, is fatal to an armies discipline. [75]

In 1853, during the serious disturbances that occurred on the Bendigo diggings, the Polish miner and military veteran Seweryn Korzelinski made some observations about the military potential of his fellow miners that emphasised how close to the mark Keegan's assessment is. Korzelinski was asked by a delegation of disgruntled miners to assist in training volunteers as an insurgent militia. He refused the request, chiding the aspiring rebels and accusing them of playing at soldiers. He warned them that after the first few hard marches and a protracted campaign, they could only fail. This was even though he conceded that if an insurgent army could win the war in one battle, there might be a chance that they would succeed, but against an assault from a trained army this was most unlikely. [76]

The insurgents mobilizing at Eureka came from the crowds who had burned Bentley's hotel, gathered in tumult on Bakery Hill to burn their licences, rioted at the Gravel Pits on 30 November, and knelt beneath the Southern Cross flag that evening to swear allegiance to each other. Despite their shared passion for their cause, they remained very much an army of individuals, and that was where their greatest weakness lay. The lack of rigid discipline and professional leadership within their ranks was not at all unexpected. What was needed was time to remedy these failings, but time to do so was not a luxury to be granted to them.

Yet, scattered through their ranks there were seeds of military potential that, if allowed time, might have blossomed into something that resembled military competence. It is certain that among the throng of insurgents there were many who had prior military service. These men would have been proficient in their personal drill and use of arms, yet unless their skills were carefully exploited, such individual prowess would not matter. It is the collective competence of the many, rather than the abilities and skills of the isolated few, that makes effective armies.

The miners themselves recognised their ineptitude. Lynch castigated his comrades for 'marching, counter-marching, jangling of military phrases, and other elementary manoeuvring of raw recruits ... no better than a see saw movement among children'.[77] He ridiculed the concentration on formal drill. In his opinion, the procurement of firearms and the training in their use was what was required. Carboni expressed his dismay at proceedings, despairing that the training consisted of 'marching, counter-marching, orders given by everybody, attended to by nobody'.[78] He also lamented that the insurgents' armed array was 'totally destitute of military knowledge'.[79] Allen, veteran of Waterloo and one who presumably could judge such matters, referred to one group he saw drilling as 'an awkward squad'.[80]

There were glimmers, however, that not all the insurgents did was a waste of time, and that not all of them were such awkward soldiers. Amos, a former soldier, described how he was impressed by the demeanour of the company of insurgents who approached his camp on the Saturday. These were the men who had automatically and universally shielded their firelocks from the rain. He mentioned that such a skill was very hard to teach to recruits, and certainly an unexpected attribute to expect from the miners. The obvious conclusion is that the men Amos saw did have some degree of military training.

Richard Allan mentioned a company of Californians and Americans who were veterans of the Mexican-American War. He referred to these men as the Californian Rifles, recalling that they possessed 'a considerable knowledge of military evolutions'.[81] This was Nelson's company of riflemen. Yet this was only one company in a force that for the most part certainly seemed to live up to Allen's description of awkwardness.

What makes an effective army is not just determined by the abilities of the soldiers who fill its ranks. Its command structure is of vital importance. If the rank and file and sub-units of an army are its body and limbs, the commanders are its brains. The decisions made and orders issued by them directly affect the army as a whole. In peacetime, this can mean that soldiers are appropriately trained, paid on time, fed or not fed, well housed or not. In wartime, leadership, good or bad, is fundamental in determining victory or defeat. It is therefore important that there be structures in place within an army to enable the directives of its leaders to be transmitted down the chain of command and understood by their recipients as they were intended to be by their originators. An army in which those structures do not exist or are dysfunctional is deeply flawed. This was unfortunately the case for the insurgents at Eureka.

There was only a veneer of a command structure within the insurgent army. Lalor had accepted the position of commander-in-chief. Vern was something of a de-facto second-in-command, at least until James McGill superseded him on the Saturday afternoon. Captains for each company were elected from among the ranks. There had also been some thought given by Vern to how the companies and sub units of armed insurgents were to be organised.[82] This was, however, as far as any concessions to an efficient military structure went. Who then, among the more notable insurgents, possessed the skills and talents necessary to provide the military leadership so desperately needed?

The insurgent council of twelve met in Martin Diamond's store, located half-in and half-out of the stockade. Accounts left by Lynch and Carboni of the members of that council provide us with something of a gauge by which we can judge their individual military potential. Not all the members of the council are discussed. Carboni deliberately declines to name or comment on five of them, except to cast aspersions against them. Lynch's silence on those he does not mention is never explained. From those who were described, the pool of military talent was, while shallow, not entirely dry.

Timothy Hayes had been elected chairman of the Ballarat Reform League and was a leading member of the insurgent council. He was a tall man who stood out in the crowd. Born in Kilkenny, Ireland he was reputed to be an authority on military engineering, an interesting fact but one that seems to have had no bearing at all on the events to occurred at Eureka.[83] Carboni eulogised Hayes, saying of him that he was a noble fellow who had 'a liberal mind ... and above all a kind heart'. Then in typical fashion, he casts a shadow over the glowing image he has just created by implying that he was lazy and preferred to give orders rather than work.[84] Lynch also mentions Hayes, describing him as enjoying pride of place at the monster meetings by common consent, and as a suave orator.[85] Hayes did not take part in the fighting at Eureka, despite being arrested and put on trial after the event. Whatever military talents he did possess never became apparent.

Thonen was another prominent member of the insurgent council. He was a 24 year old German who grew up in Elberfeld, Westphalia, a politically active town during the radical era of the 1840s. According to Carboni he was a proud man who had refused to humble himself on one occasion, by declining to stoop to pick up a shilling deliberately dropped into the dirt by a customer for his lemonade. He was also, according to Carboni, of independent character, shrewd, honest, benevolent, thoughtful and dismissive of knaves. Thonen was honest, belonging to 'that cast of men whose word is their bond'.[86] One interesting facet was that he was an excellent chess player, implying that he had a head for tactics and strategy. Significantly, he commanded one of the divisions of 'riflemen' inside the stockade during the battle. He died defending the stockade, certainly having the courage of his convictions, and he may well have possessed the necessary temperament and aptitude to possibly be an effective military leader, if he had survived.

Manning was the third member of the insurgent council mentioned by Carboni. About 40 years old, bald headed and something of an indiscreet drinker, he was a fierce defender of the gold miner's rights, and was credited with having written some of the most inflammatory of the *Star*'s diatribes against government impositions on the miners.[87] Manning's predilection for rebel causes was incontestable. Born in

Ireland, he was an advocate for Irish rights in the struggle against British domination, a stance that would in later life see him forced out of New Zealand. Both Lynch and Carboni mention his essential Irish-ness. Yet he spent his childhood and young adulthood in New York City. Manning, the idealistic, spirited 'Irishman', had his rebellious character tempered in a Yankee crucible.

He was certainly a firebrand revolutionary, being involved in the political events leading up to Eureka and playing an active role defending the stockade. He had urged an attack against the government camp on the Saturday night, a course of action that was considered, but then not taken up. Manning brought an articulate, passionate, pen to the insurgent cause, and was personally committed to doing whatever it took to remedy the plight of the miners. Yet he was not a military man. His contribution to the defence of the stockade was that of an individual, and he offered no tactical or strategic leadership. Such talents would have to be found elsewhere.

It was the German digger and prominent insurrectionist Vern who seemed to promise the best hope for military leadership. Yet what we know of him is indefinite and contradictory, much of it being from accounts left by Carboni, who detested him. However, despite Carboni's relentless mocking, Vern does seem to have possessed at least the potential for military command. He was well read on military subjects, and his military knowledge does seem to have been considerable. Unwilling to concede any such thing, Carboni referred to Vern's frequent expositions on military subjects as nothing more than, 'blabberdom'.[88] Lynch, however, never accused Vern of such a failing, recalling that he possessed a military knowledge that 'comprehended the whole system of warfare, every mode of attack and defence'.[89] At Eureka, however, Vern's knowledge did not seem to have translated into practical application.

Unfortunately Vern was a braggart, and given to frequent rhetorical bombast. This aspect of his personality seems to have impressed itself upon everyone who met or saw him. He made frequent long-winded expostulations on military affairs, indulged in heroic speeches calling for armed insurrection, and at the same time extolled the virtues of heroic death and military glory. It did not help that he was in the habit of verbally abusing those he considered to be cowards. Vern was forever promising the imminent arrival of a company of German riflemen, but this was never realised. Such behaviour, and the fact that he was a foreigner, prejudiced people against him.

Vern certainly did seem to enjoy pushing himself forward as a main player in the radical arm of the reform movement. His self-promotion was amply rewarded when the authorities mistakenly acknowledged his status as the leading insurrectionist following Eureka, with one account describing him as 'their generalissimo ... who it was said organised their forces - a man of considerable talent, daring and impetuous'.[90] Vern's profile was so great after Eureka that the authorities offered the highest initial reward, of £500, for him, so convinced were they that he was the ringleader of the rebellion.

Yet despite his bluster and bullying nature, Vern approached the task of forming an insurgent army with some energy. He wrote instructions for the organisation of the insurgents' armed companies, and was in charge of constructing the stockade. Carboni credited him with shouting 'Du Baricaden bauen'.[91] Vern could be seen everywhere, trailing a long sword after him and exhibiting a 'martial mien'.[92] He ordered an extension of the stockade across the Melbourne road and into Warrenheip Gully, and appeared to have the authority to do so.[93] Lalor offered Vern the post of second-in-command, an offer that, according to Lalor, Vern declined. Allan claimed that there were many who wanted Vern as the commander-in-chief as he was an 'old military man', but others objected because he was a foreigner. Lalor was chosen to avoid dissension in the insurgent ranks.[94]

From all accounts Vern contributed very little if anything to the management of the defence of the stockade during the battle. He was mentioned as calling out to Lalor in response to a warning from an insurgent captain, but seems to have done very little else of note.[95] That is, of course, except for fleeing the stockade, the action that has come to mark Vern's most notable contribution to the Eureka story. He is generally remembered as a craven coward , but it is entirely possible that this image was exaggerated.

Rather than flee at the first shots, as according to the Eureka legend, it appears that Vern actually fled some time into the battle. Ferguson described him as 'the first to flee', reinforcing the legend, but does make the point that this was when the 'the splinters from the timbers of the breastwork were flying the thickest'. This clearly indicates that Vern fled sometime during the battle, not at its beginning. Ferguson, whom Vern had accused of cowardice the Friday night before Eureka, noted the irony and called out after Vern, who shouted in reply that he was running 'to stop the others'.[96] Carboni recalled seeing Vern floundering across the stockade in an easterly direction just as 'a whole pack cut for Warrenheip'.[97]

Both these accounts place Vern's departure from the stockade at the moment when a large number of the defenders fled. As we shall discover, this occurred about ten minutes into the battle, exonerating him from the charge of being the first to flee. This does not excuse Vern for what he did, despite the fact that he must have stood his ground when bullets were flying thickly in the early stages of the battle.

In a crowd army, a trickle of men to the rear can soon become a flood as individuals under duress seek the imagined safety of the mob, even if that mob is fleeing. Such an eventuality is even more likely if a prominent leader is among those attempting to flee. By succumbing to his base instinct for self-survival, the amateur general Vern contributed to the collapse of resolve among those insurgents who were wavering, and saw him flee. As a recognised leader of the insurgent army, Vern had a responsibility to stand his ground. Lalor, Curtain, Ross and Thonen did so, the latter two paying with their lives. A professional military man would have done this, if for no other reason than to avoid the contagion of his own panic to his men. Vern, though, was not a military professional. He was at best a self-deluded military dilettante, a romantic amateur playing at soldiers. When confronted with the carnage and terror of war he surrendered to his primal instinct to survive at all costs and, in the company of many others, turned and fled.

Of all the original members of the insurgent council Vern appeared to possess the most promising military potential. That he failed utterly when his talents, whatever they may have been, were most needed, revealed on just what a shallow foundation rested the military aspirations of the insurgents.

Lalor was the commander-in-chief of the insurgents. He possessed formidable attributes as a leader of men. He also thought himself indispensable to the cause, going so far as to claim that a large majority of the men who were involved in the movement would leave if he were not retained as leader.[98]On his own admission, however, he was no military man. Any objective analysis of his contribution to the formation and training of the insurgent force and the role he played in the fighting inside the stockade only confirms Lalor's opinion of his own military attributes.

To say he did nothing, though, would be unfair. Lalor did give some direct orders. He ordered that arms and ammunition be collected and that the stockade be constructed. He also ordered the expulsion of a sly grog seller from the stockade. At one time during the fighting he may have given directions for the positions each of the companies should take inside the stockade.[99] In administering the insurgent force, he sensibly chose to delegate authority. He did this first by conceding day-to-day military matters to Vern, then to McGill. Lalor remained busy, and had gone to bed at midnight on the Saturday night, exhausted. His re-emergence on the Sunday morning was in the role of a fighting leader. There is no question that during this time he conducted himself with commendable courage, placing himself in an exposed position and attempting to organise the stockade's defence, before being struck down.

A wise man, Lalor preferred to allow those he thought could fill the military leadership roles to do so. Unfortunately for the insurgents' cause, there were far too few who could be entrusted with such responsibilities.

Curtain proved to be one man who could be relied on. An Irish miner, he was a member of the council, a storekeeper, and a lieutenant of pikemen. A devoted husband and father, according to Carboni he was also a deep thinker, but kept his thoughts to himself.[100] Curtain demonstrated that he was an effective leader of men when the pikemen under his command attacked the triumphant solders as they crossed the palisade during the battle. This followed the flight of a great many defenders, in a panic that appeared not to affect Curtain's men. Such stalwart behaviour in the face of extreme adversity revealed a group of men who possessed a fighting spirit and *esprit de corps* that could have only been nurtured by good leadership. For that achievement Curtain should be acknowledged.

Kennedy and Esmond were two other members of the insurgent council. The outspoken Kennedy played a prominent role in events leading up to Eureka, and had led a march of miners from the nearby goldfields at Creswick to Ballarat a few days earlier. Yet he was not present at the stockade when it was stormed, and his military attributes remain unknown. Esmond was inside the stockade, but seems to have contributed nothing of a military nature to the leadership of the insurgent cause.

Carboni was also a member of the council. Perhaps befitting a man whose greatest desire in life was to be a successful playwright, his military credentials are enigmatic. He claimed to have been a colonel in Garibaldi's army in Rome in 1848, and also that the role of commander-in-chief of the insurgent force was offered to him, but he declined. Whatever military talents he did possess were never revealed.

Perhaps because Carboni was never reluctant to use the pen to poison another's reputation, he received his own share of censure from some of his fellow insurgents. Lynch summed him up as a 'noisy parade with little actual true grit at the bottom'. He also described Carboni as a mountebank revolutionary and more of the craven than a hero, an interesting comment considering Carboni's vicious lampooning of Vern as a coward.[101] Kennedy had publicly derided Carboni as indulging in suicidal rant and for having to escort him away from the front of the stage at one meeting.[102] Carboni obviously enjoyed playing the soldier, accompanying one company of armed foreign miners during the initial mustering of the insurgents. Other than that, he seems to have contributed nothing more to the military preparations of the insurgent army. Nor did he take part in the battle, preferring the shelter of his chimney.

There were others who were not members of the insurgent council, but who were associated with their military preparations before Eureka, and involved either directly or indirectly with events when the stockade was attacked.

The first of these was Ross, a Canadian from Toronto.[103] He is mentioned in many contemporary sources as playing a prominent part in the defence of the stockade. He died from wounds received during the battle, falling, as some would have it, at the base of the pole on which flew the Southern Cross. Amos mentions him in command of a company of well-ordered insurgents and Lalor, Carboni, Lynch, Ferguson and Pierson all make specific mention of his presence during the battle. Ross was a trusted man. In company with Nelson, he led a company of men out to confront military reinforcements expected from Melbourne on the Friday night before Eureka.

Ross seems to have been a good organiser and leader of men. Ferguson describes him as a man of 'fire and spirit'.[104] Amos, whose previous military service placed him in a position to assess such things, was surprised at how proficient Ross' men were in the care of their arms when he saw them a few days before Eureka. Even if Ross was not responsible for such collective competency, or the men with him that day were not his own, he certainly did not prevent them behaving in such a manner. Unfortunately Ross, as with all the other insurgent leaders, had no time to reveal his true military potential. It does seem, however, that had he lived and the insurgent cause not been crushed, Ross might well have possessed the necessary military qualities.

Little is known about another of the insurgent captains, the American carpenter Nelson. Even his first name was not recorded. He commanded a company of armed insurgents known as the First Rifles or Californian Rifles, acknowledged as the best in the insurgent force. Among their ranks were veterans of the Mexican-American War, who knew what they were doing as soldiers. The only other ostensibly North American unit within the insurgent force was McGill's Independent California Rangers Revolver Brigade. They did not join the insurgents until the Saturday afternoon and, as the name suggests, were principally armed with revolvers.

Like Ross, Nelson was a trusted man. Together with Ross on the Friday night, then with McGill on the Saturday night, he marched his men out to intercept the military reinforcements expected from Melbourne. Waiting out in the bush at Warrenheip for military reinforcements, neither Nelson nor his men were at the stockade on the morning it was attacked. It is difficult to know just what potential Nelson had as a military leader, as he never had the opportunity to show his mettle. As with many of the personalities involved in the insurgency Nelson, despite being mentioned in the 1855 electoral roll as a miner, soon disappeared from history.

The most promising candidate for military command in the insurgent army was the 21 year-old McGill. An American businessman, George Train, who had met him, claimed that McGill was a graduate of West Point, having attended there under the initials A.R.B of 1850. Searches of the records of West Point graduates for 1850 do not reveal any using the initials A.R.B. Train also incorrectly claims that McGill was the commander-in-chief of the insurgent forces and president of the Reform League.[105] McGill may have claimed that he had attended West Point, or someone else may have done so. Whatever the truth, he certainly gave the impression that he had some military knowledge and experience.

There has also been a suggestion that McGill served as a private soldier, then corporal, and finally as a brevet second lieutenant with the U.S. 15th Infantry Regiment from April 1847 to 1849.[106] If this were so, it would have placed McGill in the thick of the Mexican-American War, as the 15th Regiment saw heavy action at the battles of Churubusco and Chapultepec. However, there are problems of chronology with this claim. McGill was said to be aged 21 at Eureka. This would have made him 14 years old as a private soldier, not unheard of at the time, but it would also mean that he was only 15 when made a brevet second lieutenant. Once again, this would not be completely impossible, but it would be unlikely in a front line unit, and suggests that the U.S. 15th Regiment's McGill may not have been the James McGill of Eureka. There is, however, another possibility.

In 1846, during the Mexican-American War, Jonathan D. Stevenson, a prominent New York lawyer, politician and colonel of militia, raised the 1st Regiment of New York Volunteers for service in California, then a Mexican possession. Stevenson aimed to recruit men of good background who were young, well educated and unmarried, a description that certainly fits what we know of McGill. The intention was that the men of his regiment would remain in California as settlers following their service. An important aspect of the recruitment for the regiment was that minors were recruited, making it feasible for Eureka's James McGill to join as a 13 year old, his age in 1846.

Records show that a Private James McGill belonged to Company E of the 1st New York. He was mentioned as having his face and hands severely burned in an explosion of a guardhouse on 9 December 1847.[107] Carboni, who describes McGill, makes no mention of any evidence of such injuries, which may have been because this was not the McGill of 1st New York, or that evidence of the injuries had faded by 1854. It is also possible that Carboni simply did not wish to mention them. Nevertheless, the intriguing prospect is that Eureka's McGill and the boy soldier of the 1st New York Volunteers may have been one and the same.

The impressions McGill left with his peers vary. Lynch, who was assigned to the Independent California Rangers Revolver Brigade despite being an Irishman, thought him to be 'a smart intelligent young fellow', but 'a very indifferent student, for his knowledge of drill did not go beyond the elementary course of a few hours'.[108] This is what one could expect from a man whose sole military experience had been as a young lad with a regiment that undertook mainly garrison and policing duties. Carboni did not disapprove of McGill, but, in his inimitable style, described him theatrical terms as having a somewhat violent disposition.[109] Carboni also implies that McGill had a well-developed sense of honour and was offended by Vern, who according to Carboni, 'never came up to scratch'.[110]

Allan speaks well of McGill.[111] Lynch, however, also accuses him rather bitterly of being absent without leave during the fighting for the stockade.[112] Vern, who had earlier accused all Americans of being cowards, also suggested that McGill had accepted a bribe of £800 and absented himself from the stockade just prior to the attack. Typically, Vern provides no evidence to support his accusation. In the same unsubstantiated way, McGill's wife in later years claimed that he absented himself from the stockade on the orders of the United States Consul, thus absolving her husband of the persistent suspicion that he was either a coward or that he deliberately betrayed the men at Eureka. The tragedy for McGill was that his absence from the stockade during the battle, which as we shall see was perfectly understandable and reasonable, has enabled those with axes to grind to do so at his expense. The unwarranted distrust of McGill, and subsequently all the Americans at Eureka, that this generated has lingered most unfairly down to the present day.

Regardless of latter day vilification, Eureka's McGill possessed the air of the gentleman in command about him. This was despite his nebulous martial credentials. Not only did he bring the promise of military skills to the insurgent cause, he brought with him somewhere between 100 and 200 men of the Independent California Rangers Revolver Brigade. The Rangers, who also included veterans of the Mexican-American War in their ranks, provided a significant boost to the insurgency's military capabilities. Lalor recognised McGill's qualities immediately, and placed him in command of the insurgent force on the Saturday night.[113] McGill's tenure as commander lasted less than one day. Yet of all the insurgent leaders, he does seem to have been the one who may have possessed the most promise of displaying military acumen. Unfortunately for the insurgent cause, he had virtually no time to exhibit whatever those talents may have been.

Confusion over just who was in command seems to have been the order of the day for the insurgents. For this their leaders can be justifiably criticised. On the Saturday night before the attack, many insurgents were of different minds. Some remained at their posts within the stockade, either out of a sense of duty or because they lived in the immediate vicinity, or lived far away. Some lay out in the bush at Warrenheip, watching and waiting for the soldiers from Melbourne. Others wandered off into the night to seek food, drink, sleep or company. There were two false alarms on the Saturday night, and after the second the number of diggers remaining at their posts had noticeably diminished.

A shopkeeper inside the stockade named Shanahan claimed that many men had left in search of grog, which may or may not have been the case.[114] Lalor believed that the men who left intended to return the same day, although he was at a loss to explain why so many had left the stockade without permission.[115] Although militarily unwise, there was nothing particularly unusual about this movement of men out of the stockade. This behaviour was perfectly consistent with accounts of militia forces throughout the ages. The Ballarat correspondent for the *Argus* reported that he had been told that the insurgents had been ordered to go home for the night, and all but 150 did just that.[116] Lynch made no mention of any such order, and criticised the failure of the leaders to halt the drift of men out of the stockade.[117] For whatever reason, nothing was done, and by early morning only about 150 remained.

This failure by Lalor and the other commanders on the spot highlighted the tenuous nature of their authority. Despite this, it seemed that there was some form of military hierarchy in operation. Amos mentions McGill ordering Ross not to bring him into the stockade as a prisoner on the Saturday afternoon, indicating that at least Ross recognised McGill as a superior authority.[118] Even so, there was no actual unity of command, abrogating one of the cardinal principles of war. Was one to listen to Lalor, the commander-in-chief who admitted he knew nothing about military matters, or Vern, parading about with his long sword shouting orders? Was it to McGill, or Ross or Thonen to whom one should pay heed? What if one leader gave an order that countermanded an order by another? Whose authority overrode whose? It is little wonder that Carboni, in a fit of exasperation, had described it all as a shambles.[119]

Ironically, when the attack came there was, at least for a short time, a degree of control and coordination imposed on the defenders by someone inside the stockade who was not one of the recognised insurgent leaders. An American referred to by Carboni as commanding the men in the rifle pits, and fighting like a tiger despite being wounded, was very prominent during the battle. His name was never recorded and he seems to have faded from the story almost immediately afterwards.[120]

Even if the insurgents had been able to cobble together a coherent command structure and something like a reasonably efficient military force, winning a victory over the army, while not entirely impossible, would be most unlikely. Even if the government camp had fallen to an attack from the insurgents, and an intention to attack it had been seriously mooted, any success would have been short lived. Some 800 soldiers, four artillery pieces and 50 mounted men were marching on Ballarat from Melbourne. The full weight of that force would have descended upon any insurgents they encountered in possession of the government camp. Without doubt their vengeance would have been terrible. The rank and file soldiers marching from Melbourne had already sworn to deal mercilessly with the insurgents, and spoke frequently of ripping them with the bayonet.[121]

An example that illustrates how difficult it is for a force of citizen militia to win a victory against a professional military enemy is the Battle of Bladensburg, where an American army comprised mostly of militia confronted a force of British regulars during the Anglo-American War of 1812-1815.

Unlike the situation in colonial Victoria, the citizens of the United States were used to the requirement for them to contribute to the defence of their nation. They did this by belonging to an organised system of civilian militia. The Constitution of the United States and the Militia Act of 1792 made it mandatory for all free, white, able-bodied, male citizens between the ages of 18 and 45 to make themselves available for service in an armed militia when required. This was known as the standing militia.

A volunteer militia supplemented the standing militia. The volunteers raised companies that dressed in uniforms of their own design, and trained and paraded on a less routine basis, often enjoying the occasion as a social experience rather than a strictly military one. The militia could be called out by the federal government, but only with the consent of the governor of the state from which they were summoned, a frequent source of friction between federal and state authorities. American militia could be required to suppress insurrection, enforce federal laws and fight foreign invaders. No such system existed in colonial Victoria, and none of the miners who had emigrated from Britain to Australia would have been required to fulfil such commitments in their previous lives.

At Bladensburg an army of between 5000 and 6000 Americans confronted a British force of 4370 that was marching on Washington,

D.C. The American force was overwhelmingly composed of militia from the state of Maryland and the District of Columbia. They had supporting artillery and were occupying defensive positions sited to block the British advance. The resulting battle lasted about one hour and was an unmitigated disaster for the Americans.

The chaos among the American militia at Bladensburg reflected all the weaknesses to be found within an amateur army. Many American officers were commanding for no reason other than their political connections, and with very few exceptions they proved not to be up to the task. Confused, contradictory and conflicting orders, insubordination, and sheer incompetence characterised the American command at all levels on that day. The artillery found themselves without appropriate ammunition at crucial moments, while soldiers were deployed into the most tactically disadvantageous formations at the very worst times.

Successful command in battle is more often than not an exercise in managing the frustration, confusion and chaos born of the unforseen. Given the nature of its army, such events played havoc in the American lines. As the British advanced they fired a barrage of noisy but ineffectual rockets. Whole battalions of American militia, half fed and with little sleep, and unsettled by days of marching back and forth for no apparent reason, panicked and fled without firing a shot.

Not all ran, however. As happened at Eureka, a significant number stood their ground. Unlike the insurgents at Eureka, they managed to repulse one British attack. This triumph was short lived, for when the British advance resumed, the militia soon found that they were being fired at from three sides. In a hopeless situation, they were ordered to run for it by their officers. The British pursued vigorously, the Americans could not rally and were swept from the field. The Bladensburg Races, as the battle became known, opened the way to Washington for the British, who burned all but one of the public buildings in the city, including the White House and Capitol.

At Bladensburg two opposing forces fought, one composed of armed citizens with a veneer of military training, and the other a hardened regular force. The militia army was the product of a community that actively fostered the militia tradition, possessed artillery, and was occupying defensive positions. The weapons, tactics, even cultural and ethnic backgrounds of both armies, were similar. The militia had the incentive of fighting to protect the capital city of their homeland against a traditional enemy. Even so, they could not stop or even seriously impede their foes. This is not to say that such an outcome was inevitable, and the militia army is always doomed to fail. With a brilliant leader, advantageous terrain, numerical superiority, an inept or overconfident enemy or a good deal of luck a militia army might prevail. None of these conditions applied at Bladensburg, nor did they at Eureka.

It is easy to criticise the leadership of the Eureka insurgency. They were, after all, ordinary men, not soldiers, who had been thrust by circumstance into positions of quasi-military authority over a boisterous mass of civilian volunteers. At Ballarat an infinitely more amateur military effort than the American model had progressed to no more than the embryonic stage. The Eureka insurgents were, as historian Geoffrey Blainey observed, very much a group that lacked the experience, and one might add the knowledge, to move from a peaceful protest to one of violence.[122] That they did manage in a very short time to gather together something like 1000 men, form recognisable military units and appoint commanders was a commendable achievement.[123] Even so, what Carboni identified as a 'confusion which baffles description' was an apt description of the army of gentlemen soldiers that had gathered together at Eureka.[124]

Chapter 3
'IN BRAVE DEVOTION TO DUTY'

We cannot appreciate why events unfolded as they did at Eureka without understanding what sort of men the soldiers and police who attacked the stockade that morning were.

Soldiers from two British army regiments, the 12th and the 40th Foot,[1] were present at the storming of the stockade. Both regiments had proud lineages and were typical of the military units scattered across the globe, garrisoning Britain's far-flung empire during the 1850s.

Known by its nickname 'The Excellers', the 40th Foot, 2nd Battalion of the Somersetshire Regiment, was originally raised in 1717 from several independent companies stationed in Newfoundland and Nova Scotia. The 40th's history over the next two centuries was typical of British regiments, participating in most of the major wars fought by Britain during the era. In 1824, the 40th landed in Australia for its first tour of duty there. During this time several companies garrisoned Van Diemen's Land, and one company was posted to New South Wales.

The first two years of the 1840s saw the 40th engaged in a hard fought campaign in Afghanistan. In December 1843, the regiment was again in action on the sub continent, where it played a notable role in a particularly hard fought battle at Maharajapore in northern India.[2] Respite in England and Ireland followed, before the regiment embarked again for Australia in July 1852. Garrison duties in Melbourne and on the major goldfields of the colony of Victoria followed. The special requirements of having to escort gold from the diggings to Melbourne led the 40th to form a mounted company, which saw a great deal of service before it was disbanded in 1857.

Some 276 men of the 40th were at Bendigo during 1853, at the time of the serious anti licence agitation that accompanied the 'Red Ribbon Rebellion'.[3] In late November 1854, a company of the 40th under the command of Captain White marched into the government camp at Ballarat. It was joined on 28 November by another contingent led by Captain Henry Wise.

The 12th Foot, East Suffolk Regiment, were known as the 'Old Dozen'. Like the 40th, the 12th had a long and proud military pedigree. Raised in 1685, the regiment had fought in most of the major wars fought by Britain in the succeeding years. In 1750 the 12th was one of the six British regiments that took part in the battle of Minden, one of the most famous actions fought by the British army during the eighteenth century. The regiment adopted a distinctive Castle and Key badge to commemorate distinguished service in Gibraltar during 1779 and 1783.

Like the 40th, the 12th saw much service in India, taking part in campaigns there from 1797 until 1809. The 1840s saw the 12th stationed in southern Africa, where it fought native enemies along the frontiers of Cape Colony. In 1852 a detachment of the 12th, and detachments from several other regiments were shipwrecked while being shipped to the Cape aboard HMS Birkenhead. As the Birkenhead sank, the soldiers were paraded on deck and the women and children aboard were allowed into the lifeboats. Only when the ship began to break up were the men permitted to fend for themselves. Until that moment, not a man stirred from his place. Fifty five men of the 12th drowned in that extraordinary example of selfless discipline.

Five companies of the 12th arrived in Australia in late 1854, two companies being sent to Ballarat in November. It was a contingent of the 12th that was ambushed by angry miners near the Eureka lead while making its way through the diggings on the night of 28 November.

Just what sort of men were these soldiers of the Queen who marched into Ballarat in the waning days of November 1854? The answer reveals a great deal about why events at Eureka turned out the way they did.

The British army of 1854 was an anachronism within the rapidly changing society of mid-nineteenth century Britain. The most rigidly hierarchical social structures and prejudices were maintained within an army dominated by an officer class that generally reflected such values. Whether from aristocratic or bourgeois backgrounds, all officers considered themselves gentlemen, a social distinction they presumed ensured a natural superiority in morals, intelligence and social standing over the common soldiers they commanded. The oft-quoted quip by the Duke of Wellington that, despite their loyalty and courage under fire, his soldiers were 'the scum of the earth', has its origin in this perception.[4] Huyghue described the men of the 12th and 40th who were forming up to march to the stockade in the early morning hours of Sunday morning as 'rude fellows'.[5] Both Wellington's and Huyghue's opinions were not far from the mark.

Within such an army, the officer's purpose within the regiment was to act as a model, of decorum and gentlemanly virtue in time of peace, and of steady, resolute, and fearless leadership in time of war. It would be commendable if an officer could master administrative detail and the intricacies of military tactics, but it was not essential that he do so. It was still considered that officers were 'fitted for command by their breeding and education' above all else.[6]

An officer's pecuniary situation was also important. An officer should preferably be a man of independent means, or at least have funds enough to live the social life expected of him.[7] Officers were also expected to purchase their commissions, ensuring that they were willing to serve the crown for honour and duty, not material reward.[8]

The system of purchase, since its inception in 1720, had been a treasured institution within the British army. Each commissioned rank had its price, and once paid became the personal property of the purchaser. Under such a system, wealth, not necessarily competence, was of crucial importance in progressing in one's career. The cost of a first commission as an ensign in a line infantry regiment was £450, increasing to £1200 in a Guards infantry regiment. Cavalry commissions would cost four times as much. £4500 would buy a lieutenant colonelcy in a line infantry regiment, with costs of £7000 and upwards for more for senior ranks.[9]

A soldier of the 40th Regiment at Eureka. A soldier of the 40th stands by the stockade in the aftermath of the battle showing the effects of the desperate close battle that has just occurred. The uniform the soldiers wore that morning was both practical and comfortable. (Original panting by Gregory Blake)

Unlikely as it might seem, such a system could produce commanders of real talent. For every swaggering martinet like the Earl of Cardigan, who paid £25,000, a gigantic sum, for the command of the 15th Hussars, and later led the Light Brigade to disaster at Balaclava in 1854,[10] there were men of outstanding talent such as Sir Charles Napier, the conqueror of Scinde in 1843, or Sir Harry Smith, the victor of Aliwal and Sobraon during the Sikh Wars.

The system was also not entirely blind to merit. Even though its primary function was to keep custodianship of the army in the hands of the moneyed classes, it was sometimes possible for men who lacked wealth or aristocratic pedigrees, but showed real talent, to be rewarded with promotion, provided, of course, that they came from the respectable classes of society.[11]

One example of this that has a particular relevance to the battle for the Eureka Stockade was Captain John Wellesley Thomas of the 40th, who commanded the attacking force. In a career that would see him retire as a lieutenant general, Thomas won his promotion from lieutenant to captain by meritorious service in India, and was promoted to major for the role he played at Eureka.

Anachronistic attitudes within the army were not just limited to class and social status. Despite a continual debate about tactical doctrine and improving efficiency within the army throughout the first half of the nineteenth century, unbending conservatism at the highest levels did much to hold back the implementation of most suggested reforms.[12] This would have the result that, apart from a few adaptations, the operational methods and small unit tactics employed by the army in 1854 were essentially unchanged from those used during the Napoleonic Wars, 40 years before.

The officers were, however, nothing without the men who filled the ranks, the common Redcoats who wielded the musket and bayonet at their superiors' orders. The men recruited into the ranks of the British army in the 1840s and 1850s tended to come from a narrow spectrum of British society. Until the mid-1840s, the Irish were the main pool of recruits for common soldiers of the British army. For well over a century, endemic hunger and poverty among the Irish peasant and labouring classes drove many young men into the army as their only alternative to lives of hopeless misery. Even a cursory glance at the muster lists of the 12th and 40th stationed in Ballarat in December 1854 reveals significant numbers of Irishmen.[13] An ironic aspect of the battle at Eureka would be the numbers of Irishmen involved on both sides.

During the 1840s, the sources of recruitment began to change. The lure of emigration to America supplanted service in the army as the more desirable vehicle for young Irishmen to satisfy their personal ambitions. By 1846 the main source of recruits had become the British urban unemployed.[14] This coincided with a transformation of British society that by 1851 saw as many people living in cities and towns as in the countryside.[15] Although life for the common soldier in 1854 could be a harsh experience, some within the ranks may have considered that it was not as harsh as the life they had escaped. The Dickensian urban wastelands endured by the British poor during the 1840s and 1850s provided fertile fields for the army's recruiting sergeants. The army at least provided steady pay, regular meals, and shelter. In the 1850s none of these things were assured to mostly illiterate, hungry, and unemployed young men from the slums of Britain's sprawling industrial cities.

Aspiring soldiers had several means of entering the service. The first was to enrol in one of the two military schools established for the purpose.[16] Another was to be born into the regiment. These children, if male, were signed onto the books, normally as a drummer boy, once they reached the appropriate age. Girls were sent away to be trained for domestic duties. Men could also fall for the entreaties of the crafty recruiting sergeant who, along with his entourage of spit and polished cronies, scoured public houses and other down and out haunts for likely recruits. There was a fourth means of enlistment that was not voluntary at all. A miscreant in court could be offered army service as the alternative to imprisonment or transportation. The motivations to enlist were as varied as the backgrounds of the recruits themselves.

Some young men did join for adventure and the lure of the uniform, but above all it was poverty that compelled men into the ranks. Daniel Defoe, the English novelist and journalist, once wrote that 'tis poverty makes men soldiers, and drives crowds into the armies'.[17] This was a sentiment echoed by Sergeant T. MacMillian, who in 1846 commented that two thirds of recruits were driven into the army by unemployment.[18]

Once having joined the colours, a recruit was inculcated into the culture of his regiment. One of the enduring traditions of the British army was, and still remains, its regimental system. Each regiment maintained its own often-unique traditions. Every man within the regiment was initiated into the ranks of what was to become almost literally his 'Mother, Sister and Mistress'.[19] Recruitment, after 1847, was for ten years, although a man would not get a pension unless he completed 22 years service.

Soldiers were paid one shilling a day, the famous 'Queen's shilling'.[20] One penny per day was added as 'beer money'. Specialists such as drummers received slightly more. Deductions were made from the soldier's pay. Four pence halfpenny was deducted for rations and eight pence halfpenny for regimental 'stoppages' and 'necessities', such as cleaning materials.[21] This enforced penury was the reality seldom explained to young men while the recruiting parties were wooing them. Soldiers who had a skill such as mending clothes, repairing boots, or carpentry, could make extra money. Serving as an officer's 'batman' or servant, and as a steward in the officer's mess or groom for an officer's horse, could also bring in extra.

Compared to contemporary wages, a soldier's pay was woefully inadequate. This was especially so in the gold inflated economic environment of Victoria. Even though soldiers received a special payment of an extra six pence per day when stationed at Ballarat, a general farm servant received 30 shillings per week with rations. A bullock driver earned 60 shillings to 70 shillings, and even a hay-cutter brought home 40 shillings per week. If a man had a trade, earnings could be as high as the £6 per week paid to blacksmiths, or £6 to £7 paid to wheelwrights.[22] At least a soldier's pay was regular, but that must have been small compensation to men earning the comparative pittance that they were.

A soldier's life was then, as it has always been, very much one of routines. Reveille was at 0500, after which, if he was in barracks, the soldier would fold his bed, clean his berth, wash himself and hang up accoutrements. Drill followed until at 0745 a bugle signalled breakfast. After breakfast more drill followed until lunch was served at noon. Afternoons consisted of more drill until 1600, when a soldier's time was his own, unless he had been detailed another task. Evening meal was at 1700, after which the soldier was free until 2000, when roll call was held. The 'Tattoo' signalling 'Lights Out' was sounded at 2200, or 2100 in winter, by the regimental band if it was present, or a bugler if not.

Barracks were almost always crowded and unsanitary. Folding metal beds and straw mattresses were issued to each man, as well as four blankets, two sheets and a barrack box in which he kept his smaller items of kit and personal belongings. The sheets were washed once a month, and the blankets only when they became excessively unclean. Barrack rooms also had a wood or coal burning stove and fuel box. There were no chairs in the barracks, and men sat on their beds. Latrines for the common soldiers were located outside the barracks. A 'urine tub' or bucket was placed in each room, for use at night or during inclement weather. As the rooms were normally poorly ventilated, the stench from these tubs was often very pronounced. The same tubs were used as the communal washing basin during the day.

Maintenance of a soldier's health in such circumstance was difficult. There was always the danger that life threatening diseases such as cholera, typhoid, tuberculosis and dysentery could very quickly spread in such a densely populated and unhygienic environment. Venereal disease also affected the soldiers' health, more than any other medical condition. In 1844, 27 percent of the 63rd Foot, The West Suffolk Regiment, were infected with either primary or secondary syphilis or gonorrhoea.[23] It was quite understandable that men who had little other opportunity for sexual satisfaction would be vulnerable to such ailments when they sought out female company from the large numbers of prostitutes who plied their trade around barracks.

Marriage was one way in which soldiers could hope to lead something like a normal family life. Soldier's wives had been part of the army since its inception, and there were strict rules and protocols to govern their behaviour and place within the regiments. It was uncommon for more than seven percent of the common soldiers to be given permission to marry. Until after the Crimean War of 1854-1856, wives lived with their husbands in the barracks. A sheet hung up provided privacy when required. When regiments were posted overseas, either on garrison or active service, some wives, drawn by lot, were allowed to accompany their husbands.[24]

The average Redcoat could not be described as a religious man, but like all soldiers he could be a reflective one in times of duress. The army's approach to catering for the spiritual needs of its men was typically one of compulsion. Religious services, known as church parades, were a weekly feature of military life. All the men of the regiment were expected to attend. At church parade the regimental chaplain, if one were present, a local clergyman if one could be found willing to do so, or the commanding officer, conducted the service.[25]

Until 1836 only clergymen of the Church of England could minister to the troops. After 1836, Roman Catholic chaplains and clergy were permitted. This would have no doubt provided some comfort to the large numbers of Irish soldiers in the ranks. Methodism, disapproved of as a subversive element within the army, was also present as something of an underground movement, with small groups of like-minded soldiers gathering in secret to sing hymns.

Chaplains had been part of the British army since the eighteenth century, but very few were to be found by the 1850s. The army that sailed to the Crimea in 1854 took with it only one chaplain. For soldiers who were sick, wounded or about to go into battle, finding spiritual counsel was problematic, or simply unavailable. An example of this occurred at Eureka, when Father Patrick Smyth, who was attempting to minister to the wounded, was ordered out of the stockade.[26]

Surprisingly, considering the generally unenlightened culture endemic within the army, it was possible for soldiers to improve their lot by education. Regimental schools had been a part of the British army since the seventeenth century. In 1811 these schools were placed on a more formal basis with the introduction of the Regimental School System. Attendance was not compulsory, and not all soldiers took advantage of the opportunity. A non-commissioned officer (NCO) had to be able to read and write, which gave some reason for an aspiring man to attend school. Education to improve literacy, however, had its limitations, with two thirds of the common soldiers remaining illiterate in 1850.[27] The comparison with the rest of the British population by 1850, in which 68 percent of males and

55 percent of females in England and Wales were literate, once again marked the army as out of step with mainstream society.[28]

It was one thing to provide schools for the soldiers, allowing them to achieve too much from the opportunity was something different. Given the army's obsession with maintaining the hierarchical status quo, an education was not allowed to encourage soldiers in any way to consider themselves the equal to their officers. Such ambitions were thought to be 'inconsistent with the habits of the country'[29]

Food has always been important for soldiers, and the Redcoat of 1854 was no exception to that rule. When in barracks, the most important part of the British common soldier's diet was meat. The type of meat depended on what was available, but generally varied between boiled meat, salt pork, or fresh beef. Bread, potatoes, coffee, tea, sugar, salt and pepper made up the rest of his fare. If a soldier wanted vegetables, he had to buy them himself or grow them in garden plots that were sometimes made available for that purpose. Food was never short, but it was not abundant either. One line from the times stated that when 'a man entered the soldier's life he should have parted with half his stomach'.[30] On long marches in the field, soldiers subsisted on biscuits, salted meat and cheese.

Alcohol, or 'grog', was a major factor in the life of the average British soldier during the nineteenth century. Soldiers found an escape from the demands of their daily routines in the bottle, and many of them became inveterate drunkards. The army issued grog to its soldiers on a regular basis, even during combat.[31] There was a belief that alcohol would not only brace the nerves of soldiers in battle, but that at other times it had a nutritional value and its regular consumption benefited the men. When local water sources were suspect, as was the case at Ballarat, grog at least provided something relatively safe to drink.

The most commonly issued grog for soldiers in 1854 was rum, and the 40th was said to have brought 140 gallons to Ballarat.[32] One of the reasons that soldiers were paid daily was the army's belief that they would binge on grog if they were paid larger lump sums. The one penny per day 'beer money' was designed to limit the amount of beer that could be purchased on a daily basis, however, it was not unknown for soldiers to form 'Boozing Schools', where they would pool their money and drink excessively. Such behaviour inevitably led to soldiers getting into trouble.

When soldiers transgressed the rules, sanctions were often severe. Striking a superior officer was a capital offence, with the assailant normally suffering death by hanging. Stoppages of pay were imposed for trivial offences, but for most other offences flogging was the prescribed punishment. Despite a widespread and growing revulsion against it within British civilian society, flogging was stubbornly maintained within the army.[33] In 1847, the maximum number of lashes a soldier could receive was reduced from 200 to 50.

Being flogged was a savage experience. The miscreant soldier was stripped to the waist, strapped to a triangle, and flogged with the 'cat o' nine tails', a whip made from nine short leather straps, the tip of each strap being studded with metal. In the infantry, a drummer delivered the blows, while in the cavalry it was the duty of a farrier. The drum major called out the strokes. Sometimes, at the discretion of the commanding officer, the delivery of the blows was timed by drum roll. This method tended to draw out the suffering. The regimental surgeon and a hospital orderly were always present, ready to intervene if the offender's life was endangered. The wounds inflicted by the 'cat' were horrific. As the unit was invariably paraded to witness punishment, they would see a salutary example of the consequences of misconduct. Flogging remained as a punishment within the British army until 1880.[34]

There were few legal escapes from service available to the common soldier. If a man fell ill or was wounded he would be discharged, sometimes with a small pension.[35] It was possible to purchase one's way out of service, and some men who had saved, or somehow procured, the required funds did do this. Constable John King purchased his release from the army in March 1853 for £18.[36] The other means of escape was desertion, an option taken by soldiers since time immemorial. Regiments like the 40th were not immune to this, with 54 soldiers deserting during 1853, and 21 doing so in 1854.[37]

Despite a period of rehabilitation during the Crimean War, the reputation of the British common soldier of the first half of the nineteenth century was generally one of a social pariah. Sir William Robertson, the tailor's son who became a field marshal, described how when he told his mother he was going to join the army, she replied 'I would rather bury you than see you in a red coat'.[38] Public houses and theatres displayed signs outside saying, 'Men in Uniform Not Admitted'.[39]

This was a widespread attitude alluded to by Rudyard Kipling in his poem *Tommy Atkins*, when a soldier seeking a pint of beer at a public house was turned away by the publican with a curt '[w]e serve no redcoats here'. Such exclusion from the mainstream of life was hardly surprising. To a great extent, the regime under which the Redcoat lived no longer shared the values of British society. Soldiers were in effect a breed apart, a very necessary breed when danger threatened, but at all other times something to be kept at arm's length.

One would think that with such a gulf existing between the civilian and military, there would have been little empathy between the miners of Eureka and the common soldiers, however, this was not necessarily so. During the build up to the Eureka rebellion, one soldier is reported to have told a group of angry miners 'never mind boys, we won't hurt you'.[40] Private George Wood of the 12th Foot helped Lalor raise funds to pursue Bentley through the courts following the murder of Scobie.[41] John O'Brien recalled that the soldiers and miners were on quite good terms before Eureka, and that a soldier named Shanahan had set fire to the stable at the Eureka Hotel during the riot there remarking as he did so that it 'might as well go up too'.[42]

Lynch recalled soldiers being 'civil and polite' to the prisoners held at the government camp following the fall of the stockade. He especially mentioned Sergeant Edward Harris of the 40th, who made repeated visits to the prisoners during the night to offer words of encouragement, sips of water and to loosen the handcuffs chafing the men's wrists.[43] Two soldiers who refused to participate in the attack on the stockade before the early morning march were arrested.[44] It does seem that, despite the gulf between the military and civilian worlds, there remained a vestige of empathy for the common folk within at least some of the soldiers.

The uniforms of the soldiers at Eureka were typical of those worn by Queen Victoria's army throughout her far-flung realms. In 1854, infantry soldiers were issued with a uniform that consisted of a long-tailed red woollen coat, dark blue trousers, black leather shoes and stiff black leather shako. Sometimes, when the weather was hot, the conditions particularly difficult, or when the men were expected to carry out special tasks, the long-tailed coat was replaced by a shorter red shell jacket, and the uncomfortable shako swapped for a much more comfortable fatigue cap. This happened at Eureka.

Private Patrick Lynott of the 40th described the uniforms that the men of the 12th and 40th were allowed to wear at Eureka as being 'as easy a uniform as we could', and specifically mentions wearing jackets.[45] Huyghue, who saw them form up before the march to Eureka, mentions the soldiers being 'bare-necked', which refers to them not having to wear their restrictive and chafing tight leather stocks beneath their collars.[46] He also depicts the soldiers at Eureka wearing shell jackets and fatigue caps in a watercolour of the fighting at the stockade that he painted some years later.[47] It was eminently sensible for the soldiers to be dressed as comfortably as possible at Eureka. The rough nature of daily life in the government camp and moving about the diggings, as well as the strength of the sun, hot wind and dryness of the Australian summer, were conducive to lighter kit and modified dress.[48]

The soldier's personal equipment consisted of a blue-painted round wooden canteen containing, at least in theory, his water. A shoulder slung haversack contained three days' rations; most typically four and half pounds (two kilograms) of cooked salted meat, as well as cheese and biscuit. There was also a knapsack that was so badly designed and uncomfortable to wear that it was the cause of numerous chest and shoulder injuries.[49] The men who assaulted the stockade most probably carried their canteens and may have carried their haversacks, but given that they were in an easy uniform and did not have to march a great distance, would certainly not have carried their knapsacks.

The style of leather cross straps worn is less certain. Prior to 1854, the normal leather equipment of a soldier consisted of two pipe-clayed white leather belts worn from each shoulder across the chest. One carried a black leather cartridge box that held 60 rounds, and a bayonet in a black, brass tipped scabbard hung from the other. Fixed at a point just below the front cross over of the two belts was a brass plate embossed with the regimental number. In 1854 waist-belts, to which were fitted a bayonet frog and small pouch for percussion caps, were introduced. These replaced the cross belt that had previously carried the bayonet. Waist-belts may have arrived in Australia in time for Eureka, and if so would probably have been worn, as they were both more practical and more comfortable than the traditional cross belts. One indicator that this may have been so is that Huyghue's watercolour depicts the soldiers wearing waist-belts.

Eureka Stockade. Samuel Huyghue's painting the fight for the stockade is surprisingly convincing considering the artistic licence he indulged in with his map of the stockade. In this painting we can clearly see the soldiers in the act of advancing against heavy fire from the stockade while mounted police work around the flank of the defenders. It is also important to note how Huyghue, who did see the soldiers march out of camp on the morning of the attack, accurately depicts the uniforms the soldiers as the more practical style rather than the more elaborate versions depicted in many other illustrations of the fighting for the stockade. (Collection: Ballarat Fine Art Gallery)

Officers' uniforms were similar to the men's, although made from superior cloth and sporting gold lace appropriate to their rank. Throughout the Victorian era officers were allowed considerable discretion in how they dressed when on campaign. One example of this was that instead of their gold-laced red coat, officers sometimes wore a shell jacket of similar style but superior cut and quality to that of the men. They could also wear an unadorned dark blue frock coat. Headdress was also discretionary, and could be anything from the regulation leather shako to a soft fatigue cap or smart peaked cap. There is no record of how the officers were dressed at Eureka, apart from Lynch's cryptic reference to a 'well accoutred officer'.[50]

Being gentlemen and expected to lead from the front, officers carried swords and pistols. While the style of swords was generally standard, the type and number of pistols carried was left to the individual. In an era when all orders were transmitted by voice or bugle, officers were expected to be easily identifiable in the heat of battle. Seeing one's officers was also very good for the morale of the soldiers. It did, however, make officers much more likely to become casualties. The fates of Wise and Lieutenant William Paul at Eureka are testimony to this vulnerability.

Unlike for officers, no discretion was allowed for the armament of the common soldiers of the 40th and 12th at Eureka. Each was armed with the Lovell's 1842 Pattern smooth bore, muzzle-loading, musket.[51] Like all smooth bore firearms, its accuracy left a lot to be desired.[52] A trained soldier could load and fire a musket three times in one minute, or up to five if pressed by circumstances and not concerned with necessarily hitting what he was aiming at.

When matters came to close quarters, the slow rate of fire of the musket could not be relied on to settle the issue. In such circumstances, the bayonet came to the fore. The bayonet in the hands of the British infantryman was a fearsome weapon. It was meant to inspire both fear in the enemy and ferocious courage in its wielder.

When using bayonets the soldiers were expected to do so without restraint. Mercy was not a consideration. The Redcoat was expected to not 'give them the false touch, but push it home to the muzzle'.[53] Even though the army did not provide its troops with standardised bayonet training at the time, many of the Redcoats at Eureka would have been no strangers to the cult of the bayonet. The miners defending the slab and cart barricade of the stockade were to learn that to their cost.

Historian Richard Holmes captured the nature of the British soldier of the 1850s succinctly when he wrote that 'we must remember they had been forged in a crucible of social change, endemic violence and economic deprivation: this harsh background bred hard men'.[54] Such were the common soldiers of the 12th and 40th Regiments who marched against the Eureka Stockade in the early morning of the 3rd of December.

Not all the troops deployed against the stockade were infantry. The mounted company of the 40th was an exception. When in full dress for ceremonial occasions, the men of this company wore a uniform based upon that of the Bombay Light Dragoons. Indeed, Carboni called these troops 'the Indian Dragoons'.[55] The uniform worn in the field was, however, very different. There, the mounted company's uniform consisted of a comfortable red jacket without facings, buff riding breeches, long black riding boots, and black shako. In summer the shako sported a white cloth cover and Havelock, a flap of cloth that provided protection from the sun to the back of the neck. Australian conditions being what they were, it is not surprising to find that even this modified uniform was adapted according to individual taste. A watercolour sketch by S.T. Gill shows an officer of the mounted 40th wearing a wide brimmed hat and smoking a pipe.[56] A contemporary painting of the attack on the stockade by Charles Doudiet shows mounted soldiers wearing similar informal headgear.[57]

The weapons and equipment carried by the mounted men of the 40th also conformed to those of the Bombay Light Dragoons. Their leather equipment consisted of a black cross belt with a cartouche pouch, and a black waist-belt with a polished metal buckle. Each enlisted man and NCO carried a muzzle-loading 1844 Pattern yeomanry carbine and a light cavalry sabre. Single shot percussion pistols were carried in saddle holsters, and revolvers, if available, were holstered in the same manner or on the waist-belt. They performed valuable service escorting gold, and were present at the Bendigo diggings during the disturbances of 1853. Thirty of these men accompanied the force that marched against the stockade.

The army formed only part of the force that attacked the Eureka Stockade. The police also played a role. Two groups of police took part in the fight. A contingent of 24 Foot Police, commanded by Carter, accompanied the infantry of the 12th and 40th who stormed the stockade. A further 70 Mounted Police (called troopers) under Sub-Inspectors James Langley, Samuel Furnell, Hussey Chomley, and Ladislaus Kossak were also present.

Of all the participants in the Eureka affair, the police are consistently the most vilified and despised. The foot police, known as *Traps*, seem to have emerged from Eureka with their reputation mostly unsullied. The most bitter vitriol was reserved for the mounted police, the *Joes* of Eureka legend.[58] Execrated as cowards who stood back during the fight, they have been written off in countless accounts of Eureka as little better than brutal murderous thugs.

The police were the nemesis of every miner on the goldfields during the early 1850s. Their frequently over-zealous enforcement of the hated miner's licence laws brought them into daily conflict with the miners. It did not help matters that some police were less than tactful in their demands for licences, and were thuggish in their application of sanctions. The suspicion of diligence born of self-interest hung heavily over them, greatly contributed to by the fact that the arresting officer could keep half of each fine imposed. Corrupt practices also besmirched the reputation of the entire force, as some police seemed to pay a disproportionate interest in the welfare of certain publicans, but actively harass others. Referring to this practice, the *Argus* castigated the police harshly for forming partnerships with various hotel owners and becoming more interested in hunting down rival grog sellers than in enforcing the law.[59]

It has been one of the most treasured articles of faith associated with the Eureka legend that the mounted police contributed nothing to the victory of the government forces. We will discover in our narrative of the battle that this is incorrect, and the police played a decisive role. Just what sort of men, then, were the police who fought at Eureka?

When gold fever struck Victoria in 1851, it affected the police as much as it did every other part of the community. The force was decimated by resignations of officers who went to seek gold.[60] The chronic lack of police that resulted from these resignations, as well as the confusing number of distinct police forces operating within the colony, made it a great challenge for those who remained to maintain law and order.[61] This was the case during 1851, and into 1853, when a great deal of the policing of the goldfields fell to the already existing Native Police. Despite performing their tasks efficiently, the vast majority of miners never accepted the aboriginal troopers of the Native Police, and they were eventually disbanded. Something had to be done, and the solution was to regulate the police and radically increase their numbers. Even this necessary reform took some time to implement.

Only on 8 January 1853 did the Victorian government pass an Act that established a single police force for the colony. This new force was still very much a work in progress, and it was not until 1856 that a Manual of Police Regulations was published. This left the management of the police on the goldfields in the hands of men who had little understanding of civil policing.

The police culture on the goldfields at the time of Eureka was that of an armed paramilitary gendarmerie. Troopers and foot police were garrisoned in centralised posts, such as the government camp at Ballarat, and not permitted to mix or fraternise with the civilian population. This created a distinct gulf between the enforcers of the law and those they were supposed to protect. The alienation of the police from the miners that this system imposed was a prime ingredient in the deterioration of relationships between the two groups on the goldfields. This was especially so when the police were compelled to enforce laws which even their senior officers thought to be unreasonable.

Some police certainly acted in a most unsavoury manner.[62] In the general environment of antipathy that had developed between miners and the police at Ballarat, it was very easy for the arbitrary, frequently brutal, and corrupt behaviour of a few individual police to malign the entire force unfairly.

In its desperate need for more men, the Victorian government hired 130 military pensioners resident in Van Diemen's Land. While no doubt there were some rough lads among these recruits, they were not ex-convicts, as the cherished myth would have us believe. By January 1853 there were 230 mounted police serving in the colony. By 1854 this had risen to 485, including a squad of nine mounted detectives.[63]

The Police at Eureka. There were mounted and foot police at Eureka. Because they were required to maintain their uniforms at their own expense the uniforms of the police were subject to the vagaries of improvisation. One suspects that after several months of service the appearance of the police became anything but uniform. Depicted here from left to right is a mounted trooper armed with an American Sharpes breech loading carbine, a mounted officer wearing an obviously very personal variation of the official uniform, and a foot policeman. These illustrations are based upon sketches found at the Victoria Police Museum. (Original Painting by Gregory Blake)

The uniforms for foot and mounted police in 1854 theoretically differed, although in practice there may have been little variation apart from the equipment carried. The mounted police uniform consisted of a dark blue shell jacket with white collar and cuffs, dark blue soft peaked cap with a white band, dark blue trousers with a double white stripe, and black spurred boots. A black leather carbine belt crossed the trooper's chest from left shoulder to right hip. This belt had a black leather cartouche box and steel ring attached from which the trooper's carbine hung.

The carbine was a muzzle-loading smooth bore model similar in every respect, except for its much shorter barrel, to the smooth bore musket carried by the infantry. Some troopers at Eureka were armed with breech loading carbines, most probably the American Sharpes breechloader fitted with the Maynard cap roll system.[64] A black waist-belt with brass buckle plate and fitting from which to hang a sabre completed the mounted trooper's accoutrements. Troopers made much use of their swords during the later stages of the fight for the stockade and the pursuit through the diggings that followed.

Depending on whether or not the trooper was armed with a revolver, a leather pistol holster could be attached to the waist-belt.[65] Uniforms were issued, but not maintained at government expense. As a result the uniforms of many police soon became ragged and subject to improvisation. A sketch in the Victoria Police Historical Unit at Melbourne shows what appears to be an officer of the mounted police wearing knee high riding boots, a shell jacket with waistcoat, a broad brimmed hat and sporting a sword and pistol. A hunting cap, similar in style to that worn by Sherlock Holmes, is shown as alternative headgear.[66]

The uniform of the foot police at Eureka is difficult to determine. Gill depicts police on foot wearing what look to be thigh length dark blue frock coats and soft peaked caps. In his sketch *Licence Inspected*, he shows a policeman dressed somewhat informally in open necked shirt, loose trousers and broad brimmed hat. The policeman carries what might be a carbine or shotgun, but wears no other belts or equipment.[67] Other sketches in the collection of the Victoria Police Historical Unit show a variety of dress styles ranging from smart shell jackets and bow ties to rather frumpy frock coats. All these figures wear the soft, peaked cap. There is no certainty that any of these uniforms were worn at Eureka. Given the nature of policing and the difficulty of maintaining uniforms, it is most likely that a combination of styles was worn. Weapons would have been muskets with bayonets, carbines with bayonets, shotguns, and a variety of pistols and revolvers. Constable John King, the foot policeman who climbed the Eureka flagpole and tore down the Eureka flag, carried a shotgun.[68]

There was no a lack of military experience among the police at Eureka. Kossak was a former cavalry soldier in the Polish insurgent army. Sub-Inspector Maurice Ximenes had served in the Spanish army during the Carlist War of the 1830s. John King was a veteran of the British army, having served in some of the most severe battles of the Sikh Wars.[69] For these men, what they experienced at Eureka was nothing new. It is certain that there would have been other military veterans among the police, especially among their military pensioners.

From all accounts the police remained under firm control during the battle for the stockade. Kossak seems to have kept the troopers under his command well in hand during the affray.[70] The foot police, in particular, were said to have conducted themselves well. Their commander, Carter, took a prominent part in the fighting inside the stockade, and was instrumental in halting shooting by soldiers into the Guard Tent, thus preventing many deaths. However, the mounted police were accused of committing heinous outrages following the fall of the stockade, for which they have been execrated, not entirely fairly, ever since. Although not a focus of this narrative, the actions of the police in the hours after the battle, while appearing brutal to civilian eyes, were perfectly consistent with the behaviour of mounted troops in pursuit of a defeated enemy, especially those considered as rebels. This would have been particularly so when those rebels had put up a much tougher fight than expected, as was the case at Eureka.

While certainly members of castes apart from mainstream society, the soldiers and police who marched to and stormed the Eureka stockade were not savages, nor were they the merciless beasts of legend. Despite the carnage that occurred as a result of a much harder battle than expected, accusations that the army and police indulged in wanton massacre are entirely inappropriate. Nonetheless, carnage did result, and the army was primarily responsible for it. For this the soldiers and police, being the instruments of coercion used by a government determined to enforce its dominance on a population that was increasingly unwilling to concede it, have been vilified for generations.

Licence Inspected. Miners have their miner's licence inspected by an armed policeman. Miners greatly resented having to comply with the incessant demands to show their miner's licences. Many of them also felt the display of arms by the police who were enforcing that law to be deeply offensive. Note the rather dishevelled appearance of the policeman. (The University of Melbourne Art Collection. Gift of the Russell and Mab Grimwade Bequest 1973.)

This was at the time, and remains, a most unfair assessment. It was only natural that when the dissent of the miners at Ballarat took on the character of an armed insurrection, the response from the government would become of necessity a primarily military one. It was in the nature of the soldier's duty, which he was fully aware was expected unquestioningly from him, that when he received orders to move against the stockade he do so without hesitation. *In brave devotion to duty* is inscribed on the memorial erected over the graves of the fallen soldiers at Ballarat. It is a sentiment entirely consistent with their actions on the day.

Chapter 4
'OUR INFATUATION'

The insurgents now had their army, and were drilling their men in armed companies. Leaders like Vern harangued those who would listen with promises to lead them to death or glory. Others attempted, with varying degrees of competence, to instil some sort of military acumen into their charges. Despite the parading, speech making and generally militaristic air, it must have been plainly obvious to those advocating a face-to-face confrontation that they could not hope to match the government's forces. Or was it?

Much of what we think we know of the plans the Eureka insurgents were making, and what their intentions were, comes from evidence presented after the event. This is an important consideration, as it has coloured much of how the story has been told in succeeding generations. It is quite understandable that some participants, keen to show themselves free from treasonous intentions, gave self-serving accounts. The reliance on these stories as the sole source of our understanding explains why the Eureka miners have been portrayed as the innocent victims of a brutal crime perpetrated by a tyrannical government.[1] While the insurgents were bloodily crushed and dispersed, it does not follow automatically that they were the aggrieved innocents of legend, or that their actual intentions were benign.

Following the events at the Gravel Pits on 30 November, there was a ground swell of support among many miners for some sort of direct action to be taken to protect themselves. The initial enthusiasm resulted in hundreds of armed men drilling. Carboni mentioned up to 800 gathered at the stockade at one time.[2] Pierson recorded in his diary seeing hundreds of armed men gathered at Bakery Hill on 30 November.[3] Allan referred to seeing 1500 men, many armed, marching to Eureka on 30 November.[4] Lynch claimed that there was a force of 700 men present on the afternoon of 2 December.[5] The police spy Goodenough confirmed that there were 500 insurgents drilling on 30 November, and that this number had risen to 1000 men on 1 December.[6] In raw numbers there were many more armed insurgents than there were soldiers and police. It must have appeared to many that they would have the numerical advantage in any conflict. And yet there did not seem to be a clear consensus among them about what path their insurgency should take, or how their armed might, such as it was, should be employed.

A debate of sorts occurred between the leading personalities. There were those such as George Black and Kennedy who advocated using the insurgent army in a purely defensive manner. Others such as Manning wished to bring on a confrontation as soon, and as dramatically, as possible. The traditional view championed by supporters of the Eureka legend is that the insurgents were dissidents who, possessing only defensive intentions, wished to argue their case from a position of armed strength, but nothing more. This, however, ignores the evidence that there was at the time an intention among many miners to take the fight to the authorities.

In hindsight it seems obvious that notions of confronting the army on its own terms were dangerous delusions. A year after Eureka, Carboni recognised this as nothing more than 'our infatuation'.[7] Despite this, there was at the time a perception among some insurgents that they could use their collective force to do something decisive.

One such plan proposed to use sheer force of numbers to attack and overrun the government camp.[8] The police spy Peters named Manning as the instigator of this plan.[9] Pierson confirmed that word had gone out that a move was to be made against the camp at ten o'clock on the Sunday morning.[10] That some insurgents took this suggestion seriously is evident from the response by Father Smyth, Kennedy and Black, who intervened on the Saturday to dissuade such aggressive plans. Rede made frequent mention of the danger of an attack on the camp in his correspondence, while the serious defensive work done at the camp in the weeks before Eureka confirmed that his concern was very real.

There were other indications that the insurgents' intentions were not benign. For two consecutive nights, on 1 and 2 December, armed groups were sent into the bush to confront expected military reinforcements from Melbourne. This was not a spontaneous, uncoordinated, event, but a deliberately considered move to ambush the army. On the first night the divisions of Nelson and Ross went out, followed on the next night by those of Nelson and McGill. The latter were by all accounts the best armed among the insurgents, and Nelson's the most capable. That it was deployed for two consecutive nights on a mission that must have resulted in armed conflict had it succeeded is significant. Carboni used the word 'intercept' to describe the intentions of these armed parties, the declared reason being to 'get at their arms and ammunition'.[11] Ferguson went one better when he declared that the object of McGill's expedition was to intercept the army 'thus weakening the government force'.[12] Both Carboni and Ferguson alluded to actions that could only have resulted in bloodshed.

Was there any hope that the insurgents lying in wait at Warrenheip could succeed in their undertaking? There were 800 soldiers, four artillery pieces, 50 sailors and 50 mounted men on the road from Melbourne. They did have a long baggage train. Pierson put it at 57 wagons, in a column a mile long.[13] Such a target might have provided some opportunity for a force attacking out of the dark to cut in and capture some carts. We do not know how many of McGill's or Nelson's men were mounted. Saddles and some horses had been requisitioned, including two taken from miners riding past the stockade, but no mention was ever made of what was done with these.[14] Nelson's men were described as riflemen, implying there were few if any horses among them. McGill's Californians may have had some horses. McGill certainly had one on the Saturday afternoon, when he rode into the stockade.[15]

What is certain is that any confrontation between the insurgents and the soldiers on the road from Melbourne would have been bloody. There could be no doubt that the soldiers would forcibly resist surrendering any of their arms, particularly the artillery. The number of casualties that would have been incurred by both sides in such circumstances can only be guessed.

If by some fluke of fortune, or rank incompetence of the soldiers, the insurgents at Warrenheip had won a victory over the reinforcement column, it could well have sown the seeds for a much wider conflict. There was a real fear in government circles that any clash which resulted in the insurgents being able to claim a success, no matter of what type, would lead to a general rebellion throughout the colony.[16] Perhaps this was the intent of some of the insurgents? Many miners were under arms at Ballarat, and expressions of support for them were evident on other goldfields.

Despite the Resident Gold Fields Commissioner at Bendigo, Joseph Panton, claiming that there was no interest in the Eureka miners' cause among the Bendigo miners, a leader of the Ballarat miners, George Holyoake, had spoken at a meeting in Bendigo, and received a 'hearty reception'.[17] One of the leaders of the Bendigo miners', William Denovan, accompanied Holyoake on his return journey to Ballarat, but they had got only as far as Castlemaine when the news of Eureka arrived.

Holyoake made speeches at Castlemaine in support the Ballarat miners. J.F. Hughes, who was there at the time, recalled that '[s]tump orators harangued the miners on their grievances, urging the raising of a contingent in Castlemaine to march to the assistance of their comrades in Ballarat'.[18] Walter Wilson, who was also present, later wrote of the events of 1854, claiming that 'had time been given or proper warning given them, that the diggers would have started "*en masse*" to assist their brethren'.[19]

Several hundred miners had begun to march from Creswick to Ballarat before Eureka, but heavy rain dampened their ardour, and only about 150 arrived. Thousands of Creswick men had, however, listened enthusiastically to Kennedy's fiery speeches earlier the same day. Howitt had observed that there might have been as many as 50,000 armed miners, one third of whom would have been mounted, available to form a guerrilla army to defend the colony against any depredations at the time by the Russians.[20] Even if only a fraction of that number took up arms in favour of a domestic political insurgency, the situation would have been catastrophic for the colonial authorities.

It is interesting that on both nights it was North American leaders, Ross the Canadian, and Nelson and McGill the Americans, who took out their men, among who were many Californians and Americans. Lying in ambush in the hope of catching a larger enemy force unawares might have appealed more to their frontier instincts than the more European revolutionary style of occupying an armed barricade and awaiting the pleasure of the enemy.

That no one among the leaders of the Eureka miners seems to have raised any objection to sending a significant number of well-armed insurgents on a mission that must have resulted in bloodshed also indicates that their intentions were not as defensive as they later claimed. There had earlier been a plan to attack the government camp, but this had been actively discouraged. The equally adventurous scheme to intercept the expected military reinforcements from Melbourne seems, however, to have been implemented without a murmur of dissent. It beggars belief that men as astute as some of those who gathered around Lalor would not have foreseen the consequences of such a direct confrontation. The only conclusion one can draw from this is that the actual intentions of a good number of the Eureka insurgents were not as pacific as legend insists, but were militarily assertive.

There were other indicators that the insurgents were intent on conflict, particularly in relation to the purpose of the stockade. Writing in 1855, Lalor went to great pains to emphasise that the intention in building the stockade was not warlike. He claimed that 'in plain truth it was nothing more than an enclosure to keep our men together, and was never erected with an eye to military defence'.[21] Numerous re-tellings of the Eureka story over the years have followed Lalor's line, emphasising the stockade's very poor, even flimsy, defensive qualities. Carboni supported this, describing it as constructed in a higgledy-piggledy manner, 'simply fenced in by a few slabs placed at random'.[22] The insurgent Allan depicted the stockade as a 'flimsy, useless construction altogether, without the slightest pretensions to military engineering, and of very little use as a defence or protection to those behind it'.[23]

Waterloo veteran Allen mentioned the hasty way the stockade was put together, emphasising that it had been constructed in three to four hours around existing tents, one of which was his. Loyal to the government and particularly disaffected with the insurgent cause, he dismissed it as 'merely slabs just merely put together higgledy-piggledy. The stockade was of no strength'.[24] Championed at the time and since by those who have wished to depict the Eureka miners as victims of disproportionate official violence, this image of the stockade has entered folklore. The reality was different.

When descriptions of the stockade given by those tasked with attacking it are considered a quite different image emerges. It was a rough but reasonably solid structure, certainly proof against musket balls, and a barrier to horsemen. While not artillery proof, it provided protection for its defenders against the personal firearms the army and police could bring to bear. The evidence for this is quite clear, and comes from the numerous accounts by both miners and the men who attacked and then stormed the stockade.

There are numerous contemporary accounts of what the stockade looked like, and even how big it was. In letters written during December 1854 Alfred Madocks, from Ipswich in England, a miner who was working a claim close to the Eureka lead, estimated that the stockade covered four acres (one and two thirds hectares).[25] Carboni described it as 'an acre of ground on the surface of a hill accessible with the greatest ease from every side'.[26] Carter testified that the stockade was 100 yards (90 metres) wide and double that in length.[27]

Various descriptions were given of its shape. A plan was produced for the State Trials in 1855 that showed it as a rectangle with one open end. When shown this plan, Amos, the Gold Commissioner posted to the Eureka diggings and quite familiar with the stockade, provided a very detailed description of it, as would be expected from a man living within a couple of hundred yards [180 metres] of the structure.[28] He was a very pedantic man, who when under oath during the trial made a point of being so exact with his answers to questions that he was pointedly accused of quibbling by the defence counsel.

Amos' pedantry, and the fact that he was formerly a soldier, are quite valuable when attempting to build a picture of the stockade. When shown the plan at the State Trials, he commented that the stockade was much stronger than was depicted. He complained that no carts were represented, these having been emplaced to support the balustrade. He went on to describe the manner in which the stockade had been constructed, testifying that the:

> slabs were placed at a great angle facing outwards; they were three or four feet separated in some places by other slabs placed crosswise, in some places by carts, and in some places by mounds of earth. The configuration of some of the slabs formed bastions. It was so strong that, although only about four feet high, there was no horse would take it at the time of the attack.[29]

Carter, who led the foot police in the battle, described the defences as 'a great number of slabs put together about breast high. A great number of them very thickly piled together'.[30] He stated that the stockade was in the shape of a parallelogram.[31] Richards recalled that the stockade 'was constructed of slabs like they use in digger's holes; they were placed across and supported themselves … They were leaning outwards'.[32] He said that the stockade had three sides, reinforcing the notion that one end was open, and that it was either a rectangle or parallelogram.[33]

Giving evidence at the State Trials, Sergeant Daniel Hegarty mentioned that the slabs 'had port-holes like', between which presumably the insurgents fired. He said that he did not notice if they did so or not. Pointing at the witness box he occupied, he mentioned that the slabs of the stockade were as high as the box, a not very useful but intriguing indicator nonetheless.[34]

When asked what was the nature of the stockade, and how it was constructed, Lynott described it as 'fenced in by slabs of wood, and the part I saw was trenched round about two feet [60 centimetres] outside the slabs of wood and, the stuff thrown over the slabs to make the fence a little firmer'. Lynott was of the opinion that musketry would not penetrate the slabs, that were about three feet (a metre) high and two or three inches (50 or 75 millimetres) thick, but then added that it 'was not a strong defence, the men jumped over it'.[35]

What Lynott meant by this must be understood from the point of view of a soldier, rather than a casual observer. As a soldier, he would have assessed the stockade from what he knew or had seen of military earthworks and entrenchments. These would often have been much more formidable defences than the stockade, hence his dismissal of its strength. His mention of a trench most probably referred to a drainage trench of no military importance, as no other accounts mention a trench of any significance. The old soldier Allen dismissed the defensive capabilities of the stockade, most likely for the same reasons as Lynott.

There are descriptions of the stockade as a 'ring'. This might have indicated a circular, oval or rhomboid shape, but could also indicate a straight-sided enclosure such as a boxing ring.[36] Huyghue drew a plan that depicted it as a semi-circle. He also left a very detailed written description. Both have been accepted by many as the definitive description of the stockade. Unfortunately for those seduced by this imaginative interpretation, everything about Huyghue's plan and description must be viewed with scepticism. Huyghue never saw the stockade. By his own admission, he viewed the devastated site of the stockade the day after it had been pulled down and burned. His interpretations would have been inspired entirely by second hand accounts that were at variance with descriptions given by those who had actually seen and fought over the stockade.

The Eureka Stockade was a much stronger defensive position than legend admits. It was a roughly built fortification of generally rectangular shape on about one and half to two acres (0.6 to 0.8 hectares) of ground. Surrounded on at least three sides by a strong fence of slabs, buttressed at intervals by piles of earth and overturned carts, it was proof against an attacker's muskets. The slab palisade was about three or four feet (one to 1.3 metres) high, enough to prevent a mounted man jumping his horse over it. Many of the slabs were angled outwards to allow men sheltering behind them to fire over the top, and possibly provide a crude *cheveaux de frise* effect.[37] Far from an inconsequential jumble of slabs, the stockade was sufficiently robust to demand deliberate and serious effort if it were to be overcome.

The actual defensive strength of the stockade was relative to the circumstances. For an attacking force lacking artillery, any continuous musket-proof obstacle that armed men could shelter behind posed a challenge, particularly if it was difficult for men on foot to cross, and mounted men could not jump it.[38] It would be easy to dismiss all these descriptions of the relative strength of the stockade as lies put forward by the servants of a government attempting to legitimise their actions, but this would be wrong. There are simply too many accounts, many given under oath, in the face of determined cross examination, that describe a structure much stronger than Lalor's flimsy fence. To suggest that the descriptions of the stockade as a much stronger structure were the product of some sinister grand conspiracy is unreasonable.

Events during the battle at Eureka also indicate that the stockade was of reasonably solid construction. The stockade provided adequate cover from the musket fire of the soldiers for a great many insurgents sheltering behind its slabs for at least ten minutes,. This could not have been possible if the stockade had been the flimsy structure of popular legend.

The stockade had its weaknesses, particularly its location. Its exact position is discussed in Chapter 6, but as Allan pointed out, the location was appalling from a defensive point of view. The stockade ran down a gentle slope, which exposed a sizeable portion of its interior to fire from nearby high ground. The approaches from the northwest, where the land was lower, also provided advantages to an attacking force.

The stockade was also far too large for the garrison remaining inside it on the Saturday night, the primary reason effective defence proved impossible when it was attacked on the Sunday morning. Despite numerous insurgents wandering off to the shelter of tents and other attractions around the diggings, a significant reason for there being too few defenders inside the stockade when the battle occurred was that several hundred of the best armed and most capable men had been sent out to Warrenheip. Their absence proved decisive.

Another aspect of the siting of the stockade that implied less than benign intentions was that it was built across the Melbourne road. This presented the government camp with a major challenge. The Melbourne road at the time was little more than a meandering swath of potholes, muddy pools and debris. It was, nevertheless, the most accessible route for wheeled traffic into and out from the Ballarat diggings, and on to Melbourne. It was also the most practical means for reinforcements and their baggage trains to reach Ballarat, an important consideration for a garrison under duress. Government officers referred to the stockade consistently as an entrenchment, such was their perception of what had been built by the insurgents. Even if an exaggeration, one could appreciate why this was their assessment for, as one historian observed, if it was 'not built for military defence, why was it built at all?'[39]

There were logical reasons why Lalor ordered the construction of the stockade. The first was the physical challenge it posed to the increasingly isolated government camp. The stockade also provided a protected base for the insurgency. Arms and ammunition could be kept there. Armed insurgents could muster there secure from raids from the mounted police or army. Weapons such as pikes could be manufactured there without interference. The insurgent council could meet in relative security behind the slab palisades. Foraging parties could range out from the stockade and return to it with requisitioned goods. It provided a convenient central point readily found by any member of the insurgency. And perhaps most important of all, a defiant insurgent flag flew boldly above the stockade, its every ripple announcing to the masses of uncommitted miners watching events that the Eureka insurgents would not be cowed.

There was no consensus among the insurgents regarding their intentions toward the government authorities at Ballarat. They had, however, built a crude, but relatively effective, fortification astride the main line of communication with Melbourne, from which they could challenge the authority of the government camp. A significant faction among them sought confrontation with the colonial authorities, and had taken steps to ensure that confrontation occurred. Despite they and their supporters loudly proclaiming in later years that they had no such intentions, the reality was that at the time, as Carboni lamented, it was very much a case of the Eureka insurgents being seduced by their militaristic infatuations.

Chapter 5
RIOT AND REVOLUTION

The Eureka insurgents may have been firm in their convictions, although not unanimous regarding their intentions, but what of those at the government camp?

In the weeks and days before Eureka, life was becoming less and less tolerable for the soldiers and police at Ballarat. Space inside the government camp had become tighter. By early December the camp garrison had risen to 450, including 150 mounted men and their horses.[1] Men had not been able to change their clothes for days, and sleep, when it could be got at all, was at a premium. Maintaining patrols of mounted vedettes and keeping a mounted force ready to respond at a moment's notice meant that troopers and mounted soldiers slept on the ground by their horses, even when it rained. Gold Commissioners could not leave the camp without being given a heavy escort.[2]

The situation was aggravated by the reluctance of contractors to supply fresh food and water to the government camp.[3] Disgruntled miners had been snapping shots at the camp for some weeks, and bullets, fired from somewhere out on the diggings, occasionally cut through the tents. Panton, the Resident Gold Commissioner of the Bendigo diggings, had visited the Ballart camp in late October, and wrote of occasional bullets passing through the canvas of the officer's mess tent during his stay.[4] Rede reported the same in a letter to Chief Commissioner of the Gold Department William Wright,[5] while Thomas, who later commanded the force that moved against the stockade, mentions shots passing over the heads of sentries.[6]

Fearing that a general assault on the camp was imminent, comprehensive plans of defence had been drawn up. Prepared and signed by Thomas, the camp defence plan of 27 October listed in methodical detail the positions and roles of all those in the camp garrison.[7] An indication of the seriousness of the situation was that the defence plan involved the preparation of firebombs to burn down the buildings surrounding the camp, lest they provide shelter to any attackers.[8] In a further sign of edginess, all civilians not resident in the streets around the camp were barred from travelling along them.[9] The solid stone structure of the Bank of Victoria, near enough to the camp to provide a threatening strongpoint if it should fall into the hands of attackers, was converted into a fortified outpost and garrisoned by armed civilians under command of police working in shifts.[10]

Despite the defensive works, there were frequent alarms. At 0400 on Saturday 2 December, word reached the camp that 400 insurgents were on Bakery Hill, close to the camp. Having heard that the insurgents had intended to attack the camp, a force of soldiers sallied out, but found nothing.[11] Those inside the camp were becoming jittery.

Rede, the son of a Royal Navy officer and 39 years old at the time, was the Resident Gold Fields Commissioner at Ballarat. He was therefore the senior government officer on the Ballarat fields, with final authority over all matters. An educated man, he had abandoned his medical studies to seek gold in Port Phillip during 1851. Rede was fortunate and did find some gold, but soon tired of the life and took up a position as a Gold Commissioner, from which he was appointed to the rank of Resident Gold Fields Commissioner at Ballarat.

Rede's attitude to his job was ambivalent. He was uncompromising in his insistence that law and order be maintained at Ballarat, especially in relation to the miner's licence law. Yet he was not entirely aloof from the concerns of the miners, and assisted them at times in practical matters, helping some to resolve their individual troubles. Carboni mentioned Rede speaking to him and one of Carboni's friends in a gentlemanly fashion, when he went to see him regarding a brick-making venture.[12]

The Government Camp Ballarat 1854. The government camp at Ballarat held a garrison of over 400 soldiers and police and was prepared to resist attack from disaffected miners in the days before Eureka. This illustration by Samuel Huyghue, who worked in the camp, shows a column of soldiers, not at all unlike that which was marching up from Melbourne at the time of Eureka, arriving at the government camp. (Collection: Ballarat Fine Art Gallery)

Rede has become the *bete noir* of the Eureka legend, and has been pilloried relentlessly as the man ultimately responsible for the carnage. In one sense this is perfectly correct. Unlike Resident Gold Fields Commissioners at places like Sofala, Bendigo and Castlemaine, Rede's regime at Ballarat was characterised by a rigidly unbending attitude to the application of the law. It was on his authority that the 'digger hunts' were pursued so relentlessly, unlicensed miners treated as common felons, and spies sent among the disaffected miners. This was despite warnings from his peers, and obvious signs of increasing unrest among the miners.

Although not corrupt himself, it was under his supervision that corruption among some government officials at Ballarat became pervasive. Most damning in the eyes of those who would see the worst in Rede was that the stockade was attacked on his authority. Rede cannot be divorced from the failure to curb corruption and the decision to move against the stockade, but he was only one actor in the tragedy that played itself out at Eureka, and working from a script not entirely of his own making. Indeed, Carboni describes Rede as a marionette whose words and movements were dictated from Toorak.[13] While exaggerated, this accusation cannot be dismissed lightly. There were other players whose direct influence on what was to occur cannot be ignored.

One of these was Pasley. The eldest son of General Sir Charles William Pasley, he was born at Chatham, England, and 30 years old at the time of Eureka. Appointed as Colonial Engineer to Victoria, he arrived in 1853, and was made an official nominee to the Legislative Council from October 1854. He held conservative views on the political future of the colony that reflected those of Hotham. Pasley was deeply concerned that that the government had too frequently given way to what he called 'popular clamour'. He predicted that if the government did so again, or the authorities at Ballarat were defeated in any physical confrontation with the miners, a general rebellion would engulf the colony.[14] For Pasley the solution was simple. It required a firm hand to put down the insurgency. Presuming that hand to be his own, he claimed to have volunteered his services and made his own way to Ballarat in late November 1854. Despite being treated with scant regard in subsequent histories of Eureka, the role Pasley played is of fundamental importance.

There was nothing especially notable about Pasley's arrival at Ballarat. Rede recorded the event simply as the arrival of an officer of engineers.[15] Such humble anonymity appears, though, to have ended very quickly with Pasley joining Rede's council of war and being appointed as *Aide-de-Camp* to Thomas. It seems that such a rapid ascension to the inner circle of decision making in the camp by a supposedly questing knight errant occurred without a murmur of protest or comment from anyone in the camp at the time. This was a very curious situation indeed.

Why would Pasley, arriving alone and apparently without anything other than his personal convictions and sense of self-importance, be so readily accepted by Rede? This is even more perplexing when it is realised that other military officers of equal rank were present in the government camp. The answer becomes obvious when one looks to Pasley's political connections. He was an appointed member of the Legislative Council, the bastion of conservative political power in the colony. As such he was a forthright champion of the existing political order, and convinced that the colony was headed for ruin if the aspirations of the miners were allowed full rein.

Pasley's role at Ballarat was the unofficial government man on the spot. His presence in Rede's council of war ensured a reliable bulwark against any diminishing of zeal. It could be argued that such an informal arrangement allowed the colonial government to distance itself from anything unsavoury that might occur. Even so, Pasley was not operating incommunicado, and began to report directly to the governor the very day he arrived in the camp, albeit through the formal channels of John Foster, the Victorian Colonial Secretary.

The tenor of those communications was consistent. Pasley was determined from the moment he arrived at Ballarat to root out and crush the dissidents on the Ballarat diggings. Penning a letter on the day he arrived, he described the insurgents as seditious, and recommended that they be put down by force. The very next day he wrote proposing a course of brinksmanship 'to bring the matter to a crisis'.[16] He profoundly abhorred what he saw to be democratic developments at Ballarat, and dismissed the miners as a clamorous mob.[17] In an ominous portent of things to come, he advised that 'the disaffected must be coerced'.[18]

Rede, on the other hand, came much more slowly to the conclusion that he had something more than a law and order problem on his hands. Before Pasley's arrival, Rede had written more that once of the need to treat unruly miners harshly, urging that they be given a fearful lesson that they would not forget. However, he always couched his advice in terms of re-establishing law and order on the diggings, not in saving the political status quo from democratic usurpers.[19]

As November wore on, Rede's gravest concerns were increasingly for the security of the government camp, as he became convinced an attack by the miners was brewing.[20] On 25 November he warned about the seditious articles being written by the *Ballarat Times*, but suggested no action in particular to deal with them. It was only on 27 November that he alluded to a political dimension to the troubles he was facing, but even then it was in response to advice he had received from Father Smyth who, in a clandestine meeting, had mentioned that among the agitators were men 'resolved to put down the government at all cost'.[21]

Even at this stage, Rede did not presume to advise his superiors in Melbourne what to do about the situation.

On 28 November, the day Pasley and military reinforcements arrived, all of this changed. The significance of 28 November should not be overlooked. From that moment, Rede seems to have undergone something of an epiphany and began to express himself in terms that mimicked Pasley. Some of this may well have been in reaction to the attack by miners on the men of the 12th Regiment that night, and Rede's fears that dark democratic forces stirred up by the Americans were at work on the diggings.[22] Writing that day, he described his plans to use the army and police to observe a meeting of miners on 29 November, and act to prevent seditious comments being made there. He also strongly advised that the editor of the *Ballarat Times* now be arrested, a noticeable hardening of attitude.

Yet, his letters still contained a note of caution and uncertainty. Unsure of just how far he could go, Rede requested in his correspondence of 29 November that legal advice be given to him on the limits of his power to act, and asked that the governor support him if he should overstep the mark.[23] On 30 November Rede described the agitation for law reform among the miners as 'merely a watchword', no doubt by that meaning a democratic conspiracy. Now echoing Pasley, he advocated that Ballarat be place under martial law, and that artillery and a strong force be sent up from Melbourne. He had become convinced that the government must crush the dissident Ballarat miners, or else risk losing all the other goldfields.[24]

On the night before Eureka Rede, with Pasley, considered the situation. A great many miners remained ambivalent about the Eureka insurgents' cause. Many were suffering economic losses because of inability to work their water sodden deep lead shafts due to the ongoing strife. Rede was very aware of this, and mentions it several times in his correspondence. This seems to have led him to a decision that the best way to diminish the attraction of insurgency to the uncommitted miners was to portray the Eureka insurgents as nothing more than selfish armed hooligans and troublemakers, intent on ruining the prosperity of all. To do this, they must be found with arms in their hands and in direct defiance of the law. If this occurred then they could be dealt with decisively without risking a popular rebellion in their support.[25] Throwing in the accusation that the insurgents were mostly foreigners would further reduce their allure to the overwhelmingly British miner population.

Despite the deteriorating physical and psychological conditions inside the camp, the soldiers and police there were compelled to maintain discipline. It was imperative that they be seen to perform their duty and maintain the law. Some thought that not to do so even for a moment might result in the potential ruin of the colony.[26] Hotham's view mirrored this exactly, when in his despatch to the Colonial Secretary in London following Eureka, he referred to a riot 'rapidly growing into a Revolution'.[27]

This was no capricious concern. For the colonial authorities of Victoria the fear was very real that there might be a serious challenge to their regime, so much that in the year following Eureka, practical steps were taken to reinforce the military garrison of Victoria. Requests were also made to despatch artillery rockets to the colony to act a deterrent to any further trouble.[28] At the root of this anxiety lay a deep-seated dread of democracy, a fear endemic within the British political establishment of the era.

It is difficult for people living in the boisterous Australian democracy of the early twenty first century to appreciate just how much at variance the concept of democracy was with 'respectable' opinion in Britain in the 1850s. British society of the mid- and late-Victorian period has been described as solemn and self satisfied.[29] It was also somewhat sanctimonious. When enunciating Britain's role in the world, Lord Palmerston, prime minister for most of the later 1850s, explained that Britain 'stand[s] at the head of moral, social and political civilisation. Our task is to lead the way and the march of other nations'.[30] For those who held such views, the threat posed to Britain's pre-eminence by democratic ideals was self-evident.

Such concerns were repeated throughout Britain during the 1850s and 1860s. Writing in January 1852, Earl Grey, the Secretary of State for the Colonies, warned Latrobe against the sense of self importance and independence that had been set loose among the people in the colony of Victoria by unexampled prosperity.[31] There was also a firm conviction throughout the era among the respectable classes in Britain that pandering to the common people would inevitably lower the tone of society. In 1861 the British liberal intellectual and historian John Acton wrote that democracy 'inevitably takes the tone of the lower portions of society'.[32] James Spence, a Liverpool businessman, in a letter to the *Times* written in 1862, passionately criticised democracy, writing that 'men move no longer by the guidance of reason, but as fish in a shoal, by the volition of the mass'.[33]

Even though written some ten years after Eureka, the London *Times* echoed such sentiments when it thundered that democracy 'were it not a failure, was a threat to ordered government … equality creates discontent as fatal as those of inequality'.[34] Given such notions, is it little wonder than Pasley dismissed the opinions of the people as clamour, and mass meetings as nothing more than the mob. In the same vein William A'Beckett, the chief prosecutor at the State Trials after Eureka, rejected the ideal of equality as 'the dream of a madman or passion of a fiend'.[35]

For those holding such opinions, it was apparent that dangerous democratic tendencies were now exhibiting themselves among the miner populations in the colony. It was essential that something be done to curb them. The dilemma was to decide how this was to be done. While the miners' reform movement remained split between those demanding physical action and those arguing for moral persuasion, it could be expected that expressions of dissent would remain rowdy but containable. All this changed on 30 November with the riot at the Gravel Pits. That event sparked the miners' armed insurgency, and from then the nature of the government response hardened. Rede, certain now that he was defending the crown against the threat of revolution, felt compelled to act and deal the insurgency a crushing blow. The means to do so were at hand. The target was obvious now. The stockade must be eliminated, and with it the looming revolution nipped in the bud.

What direct role Hotham played in the events leading up to the move against the stockade is not clear. To Hotham, a naval officer with a reputation as something of a martinet, simply to sit back and allow matters at Ballarat to run their course would be unimaginable. This was especially so given the risk that existed of what he perceived as revolution. Hotham was certainly kept informed of what was happening, receiving direct communications from Pasley. He was also communicating with Rede through enciphered messages, a method that has provided much fodder for conspiracy theorists since.[36]

From letters and enciphered notes, Hotham would have been perfectly aware of what was happening at Ballarat. He would have also been aware that, given the circumstances, he would be compelled by law to act, and in a forthright and decisive manner. The laws of the time were clear on the necessity for a colonial governor to use deadly force if necessary against any who took up arms in defiance of the authority of the crown.[37] Hotham, as governor of the colony of Victoria, was within his legal rights to use the army against the Eureka insurgents.

Despite this, and subsequent claims by those who wish to portray Hotham as a bloodstained tyrant, it is not certain that he wished undue violence to occur at Ballarat. In a note written on 1 December, the day after the Gravel Pit riot, he advised Rede to be patient and await the reinforcements despatched from Melbourne that day. Importantly, he also instructed Rede in an unambiguous manner to 'enforce the existing laws with temper, moderation and firmness'.[38] These are not the instructions of a man lusting after the insurgents' blood.

The issue of the enciphered notes does, of course, raise some degree of speculation as to what Hotham's intentions really were. Despite the best efforts of some to portray the passage of these notes as evidence that there was a dark conspiracy to wreak carnage upon the disaffected at Ballarat, it is impossible to know what passed between Rede and Hotham in them, as they no longer exist.[39] All that we do know is that at some point a decision was made to move against the stockade.

Thomas was the third member of Rede's triumvirate. He commanded the forces that stormed the Eureka Stockade. It does seem that there could have been no better man chosen to carry out that sombre and sanguinary task. Thomas was 27 years of age at Eureka. Entering the army as an ensign with the 40th Regiment in June 1839, he accompanied the 40th to Afghanistan and India, serving there during the bitter Afghan campaign of 1841-1842. Seeing serious action at Kandahar in 1842 and Maharajapore in 1843, where he was wounded, he gained valuable first hand experience of command under the most trying circumstances.[40]

Returning with the regiment to Britain, Thomas served as adjutant of the 40th from July 1846 until May 1847, a position of responsibility normally reserved for efficient and capable officers. Promoted to captain, he sailed with the regiment when it embarked for Australia, landing at Melbourne in late 1852. An officer of the 40th Light Company, in itself a mark of a man with initiative, he served with the mounted force established from picked men of the regiment.[41] Sent to Ballarat in October 1854, he soon established himself as the military authority within the government camp, preparing a detailed plan of defence for the camp. No one who knew him doubted Thomas' competence for command. Rede specifically asked for him to be given formal command of the troops at Ballarat, citing concerns about the abilities of another captain who was there at the same time.[42]

An experienced, competent and effective leader of troops, Thomas possessed a natural integrity that impressed all who met him. There is no indication that he enjoyed any extraordinary political patronage, and in all respects he appears to have been the consummate military professional. Pasley always referred to him in the most complimentary manner, being almost obsequious in his hope that Thomas would return the sentiment.[43] A correspondent for the *Argus* eulogised Thomas' qualities.[44] Even Carboni spoke well of him, relating how Thomas allowed him to go free immediately following the fall of the stockade.[45]

Only Lalor cast any aspersions against Thomas, publicly rebuking him for ordering his soldiers to fire at the Eureka insurgents without warning or provocation, a claim that we will later see was incorrect. With that exception, the general consensus was that Thomas was a fine man. It is certainly true that, despite commanding the troops who would slay so many insurgents, he escaped the approbation heaped upon others involved in the Eureka affair.

The immediate event that decided Rede's council of war to move against the dissident miners was construction of the stockade.[46]

Revenge, or hunger for military glory, as has been suggested by some, was not the motive.[47] There were practical imperatives to act quickly. Reinforcements from Melbourne were still some days away, and the efficiency of the camp's garrison, already under stress, would become increasingly problematic with each day they were not relieved. An attack on the camp by masses of armed insurgents while it was thus disadvantaged could not be ruled out. The most convenient line of communication with Melbourne had now been cut by the stockade. Most important, it appeared that an imminent revolution was brewing, and it must be put down.

One incentive to act quickly to suppress the insurrection was knowledge of a planned meeting of the council of the Ballarat Reform League, to be held on Sunday 3 December at the Adelphi Theatre. The purpose of that meeting was to elect a central committee.[48] In the volatile and militarised climate following the 30 November riot at the Gravel Pits, it was quite possible that the Physical Force faction who controlled the miners' military element could emerge dominant from that meeting. If this were permitted to happen, then the proponents of armed insurgency would enjoy a position of influence within a movement that already had well-established democratic aspirations. The Reform League's charter of 11 November was proof enough of this.

For men like Pasley, in particular, the consequences for the colony would be catastrophic if armed insurrectionists who were associated with a democratic ideology became established as the representatives of the miners at Ballarat. It was not just the guardians of the political status quo who recognised the potential dangers. In October a correspondent for the *Argus* had cautioned that the resentment among the miners could easily turn into a flame that 'kindling simultaneously in different parts of the colony, would destroy all before it'.[49] A letter written to Hotham the day after Eureka by an unnamed *Young Englishman* warned that 'a few more days, and few men on Ballarat would or could have remained longer neutral'.[50] It was in this context that Rede was required to make his decision.

Ironically Rede, the senior government officer at Ballarat, by making the decision to move against the stockade, abrogated all control over the events that followed. From the moment the military option was chosen, Thomas in every practical way assumed overall authority for suppressing the insurgency. How did he intend to do so?

Thomas's plan was not complicated. He would march his troops across the diggings under the cover of darkness, and surprise the stockade at dawn. How to get to the stockade without arousing every miner within the vicinity was the challenge.

Marching directly down the Melbourne road was out of the question. Instead, he chose to approach the stockade indirectly, using a route that would keep any observers guessing as to the intent of his force. Amos knew the ground around the stockade well. From his descriptions, Thomas could deduce the best avenues of approach. Knowing that the insurgents inside the stockade were armed, Thomas knew that it was important to find some way to gain a tactical advantage, and at the same time reduce the danger faced by his troops. To do this, he decided to use both the low and high ground near the stockade. He would first halt his force behind Stockyard Hill, a minor eminence immediately to the north of the stockade. Then he would advance against the northwest portion of the stockade from a shallow gully to the west of Stockyard Hill. By choosing to attack from lower ground, Thomas would force any insurgents who shot at his troops to fire down at them.

Before he moved, Thomas had to find a means to resolve one particular challenge. It was known that during the Saturday several hundred insurgents had been at the stockade. It would also have been known that many of these men possessed firearms, and many, especially the Americans, were quite capable of using them. If something was not done to ensure that the numbers of such men within the stockade were reduced, the success of any move against it could not be ensured. Thomas resorted to a ruse.

The nature of the deception was quite simple. It was well known that soldiers were marching from Melbourne to reinforce the garrison at Ballarat. It was not, however, known exactly when they would arrive. This uncertainty resulted in the insurgents themselves sending out two companies of armed men on the Friday night to confront the anticipated reinforcements. There were government spies among the insurgents, and Thomas would have known of this action. If the insurgents could be encouraged to do the same again on the Saturday night, it would ensure that their numbers inside the stockade were reduced when Thomas moved against them.

There are various accounts of just how this was done. O'Brien claimed that a Sergeant Brewen of the 40th personally persuaded Lalor and Black to despatch 500 of their best-armed men to Warrenheip with the intention of intercepting the approaching soldiers there. O'Brien also maintained that a Sergeant Brennan of the 40th made himself at home among the insurgents, and persuaded a number of them to go to Warrenheip. Just for good measure, he claimed that Brennan also encouraged the insurgents to waste their ammunition on target practice. [51]

As we will discover in following chapters, O'Brien's recollections need to be treated with some caution. It would make sense, however, that a government agent of some type would warn the insurgents falsely of the arrival of troops from Melbourne. Writing in his diary only a few days after the stockade fell, Pierson described how a man disguised as a miner rode out from the government camp to a position some four miles (6.5 kilometres) from the stockade, then rode in at 0300 on the Sunday, saying that 800 soldiers were only five miles (eight kilometres) away.[52] Despite Pierson's timing being unlikely, his account reinforces the notion that a false warning was delivered to the insurgents. It must, of course, be remembered that Pierson was not in the stockade, or at least never admitted to being there.

McGill and Nelson, along with several hundred of the best-armed insurgents, left the stockade sometime after midnight on the Sunday morning. Whether they did so in response to the forewarning of a government agent, or moved out on their own volition, is not certain. It may have been a combination of both, with the agent's warning merely confirming something the insurgents were intent on doing anyway. Whatever the case, the stage was set for the soldiers and police to move against the stockade.

It is quite obvious that Thomas planned his advance against the stockade in anticipation that fighting might occur. It is not certain, though, if he believed that fighting would be inevitable. Rede and Pasley had made liberal use of words such as *crush* and *fearful lesson* in their correspondence from Ballarat to Melbourne. In hindsight, they used the word *attack* when referring to what they had planned for the stockade. A great deal could be made of such words to prove beyond doubt that the government intentions for the stockade were to attack and destroy it without warning or restraint. Such an interpretation, however, ignores the fact that it was Thomas, not Rede or Pasley, who was in control of events on the ground once military action began. From all accounts, Thomas possessed a civil and considerate personality.

His orders given prior to marching on the stockade stressed that his men should take prisoner all those who did not resist, indicating that he had no intention of gratuitously unleashing the full weight of the army against the insurgents.[53] The inclusion of the magistrates Charles Hackett and George Webster in the force that marched out of the camp is most important as a pointer that an unprovoked attack was not planned. Hackett claimed later that that it was his intention to call on the insurgents to disperse, but their firing at the soldiers prevented him doing this.[54] Another sign that Thomas had not originally intended to rush unannounced at the stockade and overwhelm it was that his infantry did not have their bayonets fixed when they deployed to advance. It was not until the soldiers had closed right up to the stockade, after some ten minutes of heavy firing, that they fixed their bayonets.[55]

Regardless of what intentions Thomas might have had, it was the events that occurred in the opening moments at Eureka that dictated the subsequent dynamics of the battle. Once the first shots from the stockade had killed and wounded soldiers, any of the more measured intentions he might have had became instantly redundant. From that moment, the situation at Eureka was transformed from a police action into a full-blooded military engagement. From that moment, there could be no question of how the army responded, and the Eureka insurgents became subject to all the brutal exigencies of war. It was in such a manner that the battle for the stockade progressed inexorably to its bloody conclusion.

Chapter 6
INTO BATTLE

The moon had not set when, at about 0230 on the morning of 3 December, 276 soldiers and police began to form ranks in the gully between the government camp and Soldier's Hill. They did so without any bugle call, shouted orders, or other sound to divulge their presence.[1]

The force mustered in the gully included 87 men of the 40th under Wise, and 65 men of the 12th under Captain William Queade. There were also 30 men of the 40th Regiment's mounted company led by Lieutenant Charles Hall, and 24 police on foot under the command of Carter. Furnell, Langley, Chomley, and Kossak, with 70 mounted police, completed the muster. Thomas commanded the force, with Pasley acting as his *Aide-de-Camp*. Captain Arthur Atkinson of the 40th remained in the camp with 200 soldiers, to guard against surprise attack from the numerous insurgents known to be outside the stockade that night.[2]

On the assumption that there were traitors and spies active in the government camp, the muster was kept a profound secret.[3] Personal discipline and area security must have been well maintained, as there are no accounts suggesting that the insurgents saw or heard anything. So careful were the planners not to let their intentions slip out that Hackett, the magistrate who accompanied Thomas, was simply asked to join an 'expedition', no mention being made of what was about to occur.[4]

Secrecy had its limitations, however. The moment the orders to assemble were given, a police constable recalled that the men knew what the orders meant, and that it was obvious to them what was about to happen.[5] Even though some soldiers and police were not convinced that the insurgents should be attacked, their discipline generally held, and they stood silently waiting.[6] There would, no doubt, have been a sense of relief felt by many. The moment had finally come, when they could put an end to the incessant tension of frequent false alarms, sleepless nights, unwashed filth, and having the occasional bullet fired into their camp, which they had endured for the last five or so days.

While they waited, the infantry would have rested on their arms and the mounted men stood by their horses, probably cupping their gloved hands over their mounts' muzzles to calm them and reduce any unnecessary noise. It was claimed that the 40th took 140 gallons of rum with it to Ballarat[7] Given the Redcoats love of grog, it is almost certain that they indulged, albeit under strict supervision. But whether or not the soldiers indulged, taking their rum for a 'nip of courage', as a tonic, or just from tradition, the police had certainly enjoyed a nobbler of rum before commencing their march, an issue having been passed around their ranks.[8]

With his men gathered, Thomas addressed them, ordering them to spare insurgents who had ceased to resist.[9] He may have also ordered than any who tried to flee should be shot.[10] Then the order to march was given. The force moved off across the diggings with the soldiers in marching files, the foot police moving along each flank, and the mounted troops and police at the rear.[11] Disciplined and under perfect control, the troops moved so quietly that one who was with the column observed that a pin could almost be heard to drop.[12]

Regardless of the era, types of soldiers, or weapons employed, competent military commanders throughout history have always been compelled to address a number of fundamental factors when planning a battle. Among these are: where are the enemy; how long will it take to move one's forces to them; what will be the impact of terrain on the move; and how suitable will be the ground for the tactical deployment of one's troops once they arrive? In the lexicon of military planning, such considerations are neatly encapsulated in the phrase *Time and Space*. Thomas would have considered those factors when planning his move against the stockade.

The challenge confronting him that morning was not inconsequential. Moving troops quickly and in good order across the diggings in the moonless pre-dawn dark would not be easy. There was very little that was fixed or permanent about the diggings. The energy that drove the miners' insatiable quest for gold created a constantly shifting and confused landscape, in which artificial landmarks did not remain for long.[13]

Great patches of the Ballarat diggings in 1854 were a topographic nightmare, a maelstrom of uprooted earth, mullock heaps and piles of clay. Added to this were deep, uncovered, abandoned shafts, shallow, ankle twisting, shepherd-holes, muddy pools, discarded slabs, logs, survey pegs, guy ropes, chains and any number of other potentially hazardous entanglements and traps. Strewn randomly across the landscape were discarded bottles and shards of broken glass that presented a hazard for the movement of horses and men. All of these obstacles existed among an eternally shifting confusion of canvas tents and huts.[14]

The diggings were difficult to navigate with any confidence during daylight. At night, attempting to cross the diggings could be both disorientating and dangerous.[15] The experience of the contingent of the 12th Regiment on the night of 28 November was an example of this. At about 2100 that night, and progressing very slowly due to tired horses and unfamiliarity with the confusing surroundings, the column of carts was feeling its way forward. This uncertainty slowed, then halted, the column, allowing it to be attacked by angry miners in the vicinity of the Eureka lead. Ironically, the situation was made more of a challenge for Thomas as a result of his own actions. An order of 2 December issued on his authority, that lights could not be shown in the district surrounding the camp, would have intensified the darkness on the diggings.[16]

Even so, it made very good tactical sense to commence the move against the Eureka insurgents in as near to complete darkness as possible. The intention was for the soldiers to arrive at the stockade as a surprise, and only darkness could facilitate this. Therefore, timings needed to be carefully considered. The moon over Ballarat on the night of 2-3 December 1854 was bright, making it highly likely that any movement under it would be noticed.[17] An experienced soldier, Thomas would have been fully aware of the need to wait for the moon to set.

There are many conflicting estimates of when Thomas' force commenced its march.[18] There is, however, one account of the departure that allows us to fix a specific time. Huyghue was asleep in his cot inside the government camp when awakened by a friend. Grabbing up his arms, he left his tent only to hear 'the muffled tramp of the soldiery who had moved down the gully to the northward'. Then, most importantly, he added that the 'moon had now gone down and the stars were twinkling coldly in the grey approach of dawn'.[19]

This enables us to pinpoint to within a few minutes the time Thomas' force began to move. On the morning of Sunday 3 December 1854, the moon set at Ballarat at 0352. The sky would have begun to noticeably lighten from the onset of civil twilight at 0427, and full sunrise occurred at 0457.[20] When these times are compared to known events at the beginning of the battle, Huyghue's observation allows us to fix a time for the commencement of Thomas' march, and it can be deduced how much time he had available to complete it.

Lynch, sheltering behind the slab palisade of the stockade, wrote that the first warning of the soldiers' presence was sounded 'just as the grey dawn was breaking in the eastern sky'.[21] This places the time at, or very close to, the onset of civil twilight, which occurred at 0427. Accepting Huyghue's reported departure time for the soldiers and police as sometime about 0352, and accepting that there were certainly variables as to how precise Huyghue was in his estimate of when the moon had set, this allows perhaps up to 50 minutes to cover the distance between the camp and the stockade. Included in this time would also be the necessity to deploy into formation, and advance before the first shot was fired at about 0435.

Just exactly where did Thomas's men march that morning, given the time they had available? There has been perennial debate since the event about the exact site of the Eureka Stockade. Even in the 1880s, when many who were at Eureka were still alive, no agreement could be reached.[22] Contradictory claims are still argued passionately, and no consensus has been achieved. It has been claimed that the site of the stockade has been 'lost', and that it remains for the chance discovery of some long lost document to solve the mystery of where it actually was.[23] Such a claim is far too pessimistic. Using the military logic of time and space, it is possible to estimate the stockade's location quite convincingly.

Operations of War, published in 1914, when soldiers still usually moved on foot or horseback, described the speed that troops could be expected to march when on good roads. Its author, General Sir Edward Hamley, explained that small detachments could be expected to average three miles (five kilometres) per hour, including some halts of a few minutes. Over the short distances at Eureka, the rate could have been marginally faster. Hamley warned that the rate could be 'much affected' by adverse conditions such as heat, strong winds, snow, mud, steep hills or darkness.[24] The direct relevance to Eureka of his reference to difficult terrain and darkness limiting the rate at which soldiers were able to march is obvious. He does not include the necessity to move in silence, which would have been another factor limiting the rate at which Thomas' men could move. The *Field Service Pocket Book* of 1914, used by British and Empire officers during the First World War to plan operations, follows Hamley's estimates.

That part of Thomas' force was mounted would have made no difference to the speed the troops moved. It is an axiom that military formations tend to move at the speed of their slowest elements. The integrity of a unit is always considered preferable to disunity and haste when on the move, particularly in circumstances where joint action is required at the destination. It would also have been counterproductive for the mounted part of the force to arrive early, ruining any chance of surprise before the more combat-capable infantry could be deployed. Two accounts say that the mounted element of the force took up the rear of the marching column.[25] Such an order of march would have been consistent with the good management of a force moving through difficult and potentially hostile terrain at night, and to be expected from an officer of Thomas' experience.

The only contemporary map of the route the soldiers took from the camp to the stockade is the plan produced in 1855 for the State Trials held after Eureka.[26] It shows the site of the Eureka Stockade located beneath Black Hill, about a mile (1.5 kilometres) northwest of the currently accepted site. The general route taken by Thomas' force depicted on this plan, as opposed to its exact measurements, has never been credibly disputed. The same cannot be said for the plan's depiction of where the stockade was.[27] It is worthwhile digressing, to investigate just how far the force could have moved at differing speeds. This will give a guide to the distances they could have covered, and help estimate to where they were marching.

Marching rates for Thomas' men would vary between an optimal rate of three miles (five kilometres) per hour down to two miles (a little over three kilometres) per hour, depending on the conditions they encountered. If they began their march later, this would allow them to cover roughly 2200 yards (2 kilometres), while if they started earlier they could have covered around 4000 yards (some 3.5 kilometres).

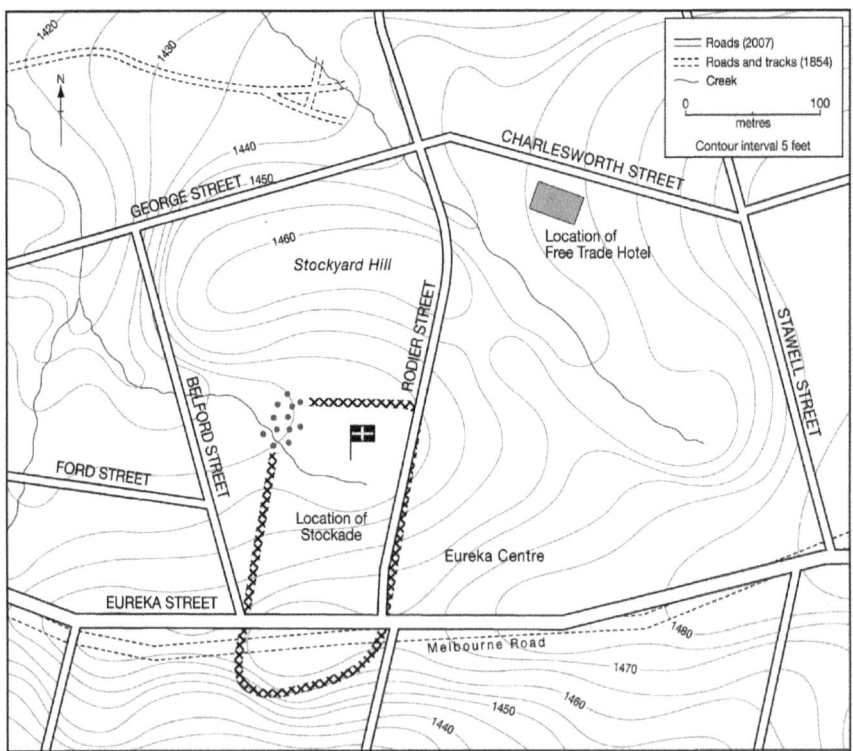

The location of the stockade. A careful consideration of the battle for the stockade indicates that the actual site of the stockade was to the west of Rodier street and the Eureka centre, generally conforming to the line of the modern Belford Street.

In themselves, these figures mean nothing until we find some ground to which we can relate them. When the State Trial plan's site for the stockade is compared with the terrain, it becomes apparent that there is no ground in the immediate area depicted in the plan even remotely resembling that described in accounts of the battle. We must therefore look elsewhere for the site.

The *1858 Geological Survey Map of Ballaarat* enables us to study in detail the terrain of the Ballarat diggings as it was in the mid 1850s. Printed in colour, the map uses the archaic scale of one inch equals three chains, and has contour lines drawn at five foot (1.5 metre) intervals. This map is invaluable in helping identify which terrain on the diggings fits tactical descriptions of the Eureka battle.[28] When this is done, the only matching ground is found near, but further to the west from, that currently popularised as the site of the stockade, and partially occupied by the modern Eureka Centre.[29] We will call this tactically appropriate ground the *preferred site*. On a map of modern Ballarat, it lies in the area in East Ballarat bounded by Eureka Street to the north, and running towards the southeast down sloping ground along the line of Belford Street. This is essentially the same site as the one identified by researchers from the Ballarat Gold Museum, which has recorded its findings in a very detailed but unpublished manuscript.[30]

The ground at the preferred site includes many features named in accounts of Eureka. The most prominent of these is Stockyard Hill, a low feature now somewhat truncated by excavation and now obscured beneath houses and trees. When standing on the modern Eureka Street, which was the Old Melbourne Road, a steady gentle slope is evident leading down to the base of Stockyard Hill, matching descriptions of that sector of the stockade as the 'lower part'.[31] On the other side of Eureka Street, behind an observer looking at Stockyard Hill, the ground drops away quite steeply, providing the cover which some mounted troopers took advantage of during the battle.[32]

Looking back towards Stockyard Hill, and to the left of the preferred site, the ground drops away, beginning to dip noticeably at about 75 yards (around 70 metres) into what would have been the gully from which Thomas began his advance on the stockade. This feature of the ground is most obvious when standing at the intersection of Belford and Ford streets, which lies about 100 yards (90 metres) down Belford Street from the rise on which Eureka Street sits. This would have been the lower part of the stockade spoken of by Carboni, and near where the Californians' rifle pits were. About 200 yards (180 metres) to the northeast is the gully through which the troops marched, on the eastern slopes of which the Free Trade Hotel possibly stood.

The most crucial indicator that the preferred site is indeed that of the stockade relates to the position of Stockyard Hill. Henry Gyles Turner in *Our Little Rebellion* mentions that soldiers posted themselves on Stockyard Hill, so that they could fire down into the stockade in support of the attacking troops.[33] Richards confirms that soldiers were on Stockyard Hill, and they therefore may have used it as a base from which to provide supporting fire.[34] While one could argue with the number of soldiers Turner says were deployed on the hill, there is no reason to dismiss his claim that something of this nature occurred. It makes good military sense to have placed men on Stockyard Hill, which dominated the stockade. From there they could fire in support of the advancing troops. How this could have been done effectively with the weapons the soldiers were using on the day is the question, the answer to which reveals the strongest argument yet for the most likely site of the stockade.

The Redcoats at Eureka were armed with the 1842 Pattern Lovell smooth bore musket. This was a notoriously inaccurate weapon with a very limited effective range, described by one officer of the era as a weapon that 'will throw a ball with some certainty about the distance a man can throw a stone'.[35] Napier, who led British troops to victory in India, was under no illusions about the weapon with which his soldiers were armed, stating that the 'short range and very uncertain flight of shot from the musket begets the necessity of closing with the enemy'.[36]

The optimum effective range for any smooth bore musket was under 100 yards (90 metres), but the closer the better. Hitting anything aimed at beyond this distance was unlikely, and firing at targets 150 yards (135 metres) or further away was a fruitless exercise. One British colonel lamented that men firing at such ranges might as well fire at the moon.[37] Even if they did strike home, many musket balls at such ranges would not retain enough force to inflict serious injuries or death. Such spent balls, as they were called, had so little penetrating power that the silk handkerchief of one officer stopped several of them during the battle of Vitoria in Spain during 1814. Another described being hit by a spent ball as being like a rap from a stick or tap from a fan.[38]

For Thomas to order his men to fire at ranges beyond 100 yards (90 metres) in support of his advance would simply have been ordering them to waste ammunition. He would not have contemplated such a thing. Even at 100 yards, the majority of shots would be guaranteed to miss. Troops firing from Stockyard Hill in support of the advance against the stockade would certainly have been no further away. As the advancing infantry skirted to the west of Stockyard Hill and then moved against the northwest corner of the stockade, this would place any troops firing in support somewhere on the western portion of Stockyard Hill. Posting men anywhere further to the east along the length of the hill, from where they could not hope to influence the fight with their fire, would have been pointless. Places further to the east of the preferred site, including the modern Eureka Centre, more than 150 yards (135 metres away) cannot be the location for the stockade. Only the preferred site complies with the ballistic tyranny imposed by the muskets used on the day.

Determining the most probable site for the stockade produces a serious conundrum. How could the soldiers and police who marched out of the government camp at about the time the moon set have covered the distance from the camp to the preferred site in the time available to them? Once again we must fall back upon what was and what was not possible in the circumstances.

Tracing the course of the route taken from the camp to a position behind Stockyard Hill on the *1858 Geological Survey Map of Ballaarat* gives us a distance of approximately 3200 yards (2900 metres). Even with the reduced marching speeds imposed on Thomas' force by the restrictions imposed by night and difficult terrain, there is evidence to suggest that the ground from the camp to the stockade was covered relatively quickly.

Thomas stated that his force arrived 'in about half an hour, in front of the entrenchment, and about 300 yards from it'.[39] Amos recalled that it took a half to three quarters of an hour to reach the stockade.[40] Huyghue estimated that the firing at the stockade began about three quarters of an hour after the troops departed the camp.[41] The correlation of these estimates, and the period between the setting of the moon and the onset of civil twilight, is clearly evident.

Other accounts from those who participated in the march also indicate the short time taken to reach the stockade. Private John Neill of the 40th recalled that the soldiers' march had not gone 300 yards (270 metres) before a shot was fired from Black Hill to warn the stockade of their approach.[42] Lynott, from the same regiment, stated that the warning shot from Black Hill was followed two minutes later by another shot. He said that when this occurred, the soldiers were in the shelter of some rising ground only a couple of hundred yards from the stockade, by which he quite obviously meant Stockyard Hill.[43] Even if both Neill and Lynott were confused about the exact passage of time before they heard the shots, their accounts highlight their perception of the short time from the beginning of the march until they halted.

In another piece of evidence that supports a more rapid approach of the soldiers and police than expected, Lalor recalled when he first heard of their approach. Writing to the *Age* on 10 August 1855, he stated 'I believe that one or two signal shots were fired by our sentries … we would have retreated but it was then too late'.[44] Those one or two signal shots sound very much like those mentioned by Neill and Lynott. Lalor's account of what he heard indicates that the shots were fired prior to the action, not its opening shots, and thus occurred sometime before civil twilight at 0427, during the period the soldiers and police were approaching. If Lalor thought it was too late to retreat when the shots were heard, it implies that they were moving very expeditiously as they crossed the diggings.

Having established that the preferred site is the most likely site of the stockade, how can we resolve the timing for the move from the camp to the stockade? Every aspect of the terrain and other physical conditions would seem to have conspired to delay the soldiers and police, yet they crossed the ground in a relatively short time. There must have been some way that their marching rate could have been increased. What could that have been?

Walking tracks existed on the Ballarat diggings. Miners required somewhere safe to walk and wheel carts, barrows and all the other impedimenta of their trade. A painting by the artist Eugene von Guerard depicting the Ballarat diggings in 1854 clearly shows such tracks, and a sketch of the Eureka Lead in 1853 depicts a track wide enough to ride a horse on.[45]

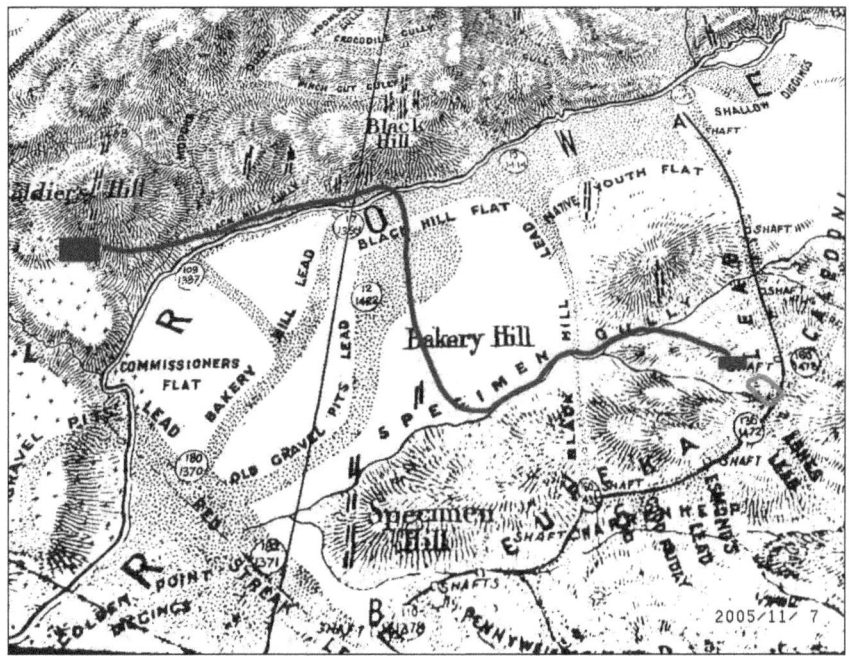

The route marched by the soldiers and police to the stockade. The exact route marched is unknown but this map indicates the general route that was taken by Thomas's force as it marched to the stockade in the early hours of Sunday morning 3 December 1854.

A map produced as part of the work done by the Ballarat Gold Museum to locate the site of the stockade shows such tracks. While they are the work of a modern investigator, they reflect the results of considerable research.[46] Anecdotal evidence from those who lived among the old diggings in more modern times also confirms the presence of tracks.[47] Logical deduction, anecdote, contemporary visual representations, and the maps prepared by the Ballarat Gold Museum all lead to the inescapable conclusion that there were tracks crossing the Ballarat diggings at the time of Eureka. It was by using these tracks that Thomas was able to move his troops as quickly as he apparently did.

Two tracks shown on the Gold Museum's map are of particular importance. One runs from somewhere northwest of the stockade, emerging onto the map in the direction of the government camp, and following a southerly course. When testifying at the State Trials, Webster described the route taken that night, saying that they 'went close by the Flat in front of Black Hill, and came in sight of the stockade about 200 yards off'.[48] There is a flat to the south of Black Hill. Did 'close by' mean that the force skirted the flat to the west? If so, the course he indicated would follow the route of the southbound track indicated on the Gold Museum map.

When the soldiers and police marched from the camp they headed north for a short distance, crossed Yarowee Creek, then turned south, recrossing the creek. After continuing south, they made an abrupt turn to the east. From there, they proceeded to the rear of Stockyard Hill, in the immediate vicinity of the stockade. The tracks identified by the Ballarat Gold Museum show that as the southbound track is followed, there comes a point were another track branches eastward at right angles. This branch takes a course directly to the site of the Eureka Hotel, then runs along the bank of Specimen Gully Creek and towards the vicinity of Stockyard Hill. The similarity to the route taken by the soldiers in the march to the stockade is noticeable.

A further sign that an easterly track existed is the presence of a 'New Bakery Hill Road' marked on the *1858 Geological Survey map of Ballaarat*. This suggests that there must at some time have been an Old Bakery Hill Road, which would probably have been little more than a track. New Bakery Hill Road runs in a generally easterly direction, traversing the diggings at a point about where Thomas' column would have made its turn east. The track along Specimen Gully and on to Stockyard Hill marked on the Gold Museum map is an extension of what may well have been the Old Bakery Hill Road.

Private Thomas Carruthers of the mounted 40th stated that there was a road from the camp to the stockade, but that it was not direct.[49]

He may have been referring to the Melbourne road, but this was not likely, as it ran directly to the stockade, and the soldiers did not use it on their march out. His road could well have been a track, or series of tracks, running at angles to each other, hence his description of the road not being direct. It must be assumed that such tracks would have been clear of significant obstacles and wide enough to move formed bodies of men along. Given the time it appears Thomas' force took to arrive at the site of the stockade, the uncompromising military logic of time and space dictates that the soldiers and police must have used tracks to cross the goldfield.

The existence of tracks, however, did not automatically ensure a swift passage for Thomas' force. It would still have been very difficult for anyone to successfully negotiate the walking tracks in the dark. A guide who knew the ground well would be an essential requirement. Luckily for Thomas, Amos, who knew the lie of the land in the immediate area of Eureka, was available and eager to fill that role.

Thomas' men moved quickly through the pre-dawn dark, the infantry leading and mounted men bringing up the rear. The officers rode among the troops, no doubt quietly encouraging them as they trudged quickly along the tracks. The men had by now guessed where they were headed, and although they did not know what to expect when they got there, they would certainly have known what would be expected of them. Even so, their pace did not slacken and they did not get out of hand. Stepping out, they would have rested their heavy muskets on their shoulders and, head down, trusted the path blazed in the inky blackness by the man to their immediate front. It would have been implicit among them that their officers knew best, and that they were embarking on a task that was as essential as it was necessary. In a short time those they held responsible for the discomfort, frustrations and humiliation of the last few weeks were finally going to be brought to account.

Chapter 7

'FORTIETH ARE YOU GOING TO RETREAT?'

Amos guided the soldiers and police as they marched through the early morning darkness. The men moved well, following the tracks, silent and disciplined. They had not been marching long when the warning shot recalled by Neill rang out. Knowing they were now seen, Thomas turned to encourage and caution those men in earshot 'Forward and steady men! Don't fire; let the insurgents fire first. You wait for the sound of the bugle'.[1] A few minutes later, the second shot remembered by Lynott thumped skyward somewhere out in the darkness.

In the distance a bugle sounded. It seemed to Lynott that it came from within the stockade, which by that time was a short distance in front. He thought he could see men running about there.[2]

The insurgents' outpost sentries had not been caught unawares that morning. Others also knew of the soldiers' movements. Thomas Budden, a Canadian miner and friend of Ross, made his way into the stockade to warn his friend and plead with him to abandon his post.[3] Ross would have none of it, and Budden left to wait out the coming conflict near the London Hotel.[4]

As Thomas' men came closer, the insurgent sentries came running back, shouting a warning.[5] 'To Arms' resounded inside the stockade.[6] There had been at least two false alarms during the night, but this time it was in earnest. Lynch reported 'a terrible effervescence of hurry-skurry' as insurgents ran to their posts.[7]

The soldiers and police moved into position behind Stockyard Hill, about 300 yards (270 metres) from the stockade. To the right of the infantry, Furnell's mounted police, guided by Amos, began to feel their way in the dark out to towards the western face of the stockade. Mounted troops and police under Lieutenant John Hall and Langley moved a short distance to the left, still protected by rising ground to their front.

The infantry formed into close order column, a tightly packed, sensible, formation. At least some of them advanced to the top of Stockyard Hill.[8] There they waited in the dark, some maybe whispering a rosary, others no doubt thinking only of the grog that awaited them back at the camp once this duty was finished.[9] Officers and sergeants would have moved quietly among the ranks, perhaps no longer whispering to their men now that the insurgents over the hill knew something was afoot. Then, as the sky grew noticeably lighter to the east, the order came, passed along from officer to officer. The Redcoats began to advance.

Moving to the right of Stockyard Hill, the column began to ascend a shallow gully that led directly up to the stockade. Carter's foot police accompanied the leading files of infantry, who had begun to extend their formation into skirmish lines.[10] Those soldiers in the column behind the skirmishers would have slowed their advance, holding back to allow the skirmishers to perform their task. Others had already moved to the top of Stockyard Hill.[11]

Thomas, Richards, and Webster, all mounted, moved to the right of the infantry line, so that contact could be maintained with the troopers moving out in that direction. Pasley, on horseback, took post in the centre of the skirmish line.[12] Thomas mentioned deploying his men into skirmish order, with each contingent 'having its proper support'.[13] By 'support', he meant a reserve of soldiers held back in open order, a tighter formation than that of the skirmishers. This reserve would be available to respond to any unforseen eventuality, such as a sudden attack from an unexpected direction, or to overcome a stiffer resistance than expected. The reserve could also be brought up quickly to plug gaps in the skirmish line, or reinforce a threatened point. Some of this reserve was posted on Stockyard Hill to provide covering fire for the main assault.[14] It was the skirmishers, however, who led the advance against the stockade.

In 1854 the British army was employing a system of skirmishing based on the *1798 Regulations for the Exercise and Conduct of Rifles and Light Infantry on Parade and in the Field*.[15] Despite their antiquity, these regulations formed the basis of skirmishing tactics until the late nineteenth century. The skirmishers spread out in the dark, making sure to leave at least two paces between each man, although this could vary at the discretion of the officer in charge. The officers took their posts at the rear of the skirmish line. About 50 paces behind the skirmishers was the line of soldiers in support.

The numbers of skirmishers were not fixed, but determined according to need by the commander, and could be anything from a quarter to half of the total force. The numbers Thomas deployed into skirmish order and retained in reserve are not clear. The figure of 40 skirmishers is used in many texts, presumably based on that figure being given by Henry Turner in *Our Little Rebellion*, in which he also erroneously called the skirmishers the assault group, a mistake copied by many others since. Turner claimed to have based his account on official government sources, yet gives no specific details of where he uncovered the figure. Whatever the numbers actually were, Thomas deployed his skirmishers, and sent them forward under Pasley's command. Carter's foot police moved up to the right of the infantry skirmish line.

To the left rear of the infantry line, 30 mounted troopers under Langley and 30 mounted 40th under Hall deployed. Langley's and Hall's men threatened the eastern face of the stockade, and also screened the attacking force from any surprise arrivals by armed insurgents from that direction. Thomas was aware that he was facing only a fraction of the insurgent force. He hoped the best armed and more numerous component was some distance away, waiting in vain for the arrival of the reinforcements marching from Melbourne. If he was wrong, it could prove fatal to his plans. By deploying Langley and Hall as he did, Thomas avoided leaving this potential threat to chance.

Sometime between 0420 and 0430, with his men in place and the enemy to his front, Thomas ordered the advance. The skirmishers of the 12th and 40th moved forward, each man no doubt watching those to his right and left to ensure he maintained his spacing. Unlike the soldiers, who were out in the open, the sun rising behind trees to the rear of the stockade cast shadows over the insurgents' position, making it difficult to see them.[16]

To the stockade's defenders, the infantry were at first little more than indistinct shapes, mere shadows in the greying light. Lynch recounted how, as he waited behind his reinforced cover of slabs, he could 'hardly discern the military force at first'.[17] This changed as they drew closer, and the light improved, until at 150 yards (135 metres) they could be clearly seen. About this time a shot rang out from within the stockade, the first in the brief but bloody encounter at Eureka.

California Ranger Ferguson and his friend Walter Hall were at the barricade. They had been waiting in line with other Rangers inside the stockade when the warning that the soldiers were advancing had been given. The call had gone out, 'California Rangers to the front!', and they had rushed to the barricade.[18] Another American defender of the stockade, Philadelphian George Hartley, recalled the call to the fence, remembering fellow Americans Bill Melody and Burnette as the first to get there.[19]

Ferguson could clearly see the soldiers advancing. One in particular, whom he identified as an officer, attracted attention as he gave orders to his men. The insurgents had been ordered to shoot at the officers if the army attacked the stockade.[20] He watched as Burnette, described by stockade defender Henry Sutherland as 'a little fellow formerly a barber'[21] stepped forward, lifted his rifle, and fired what Ferguson called the first shot of the 'Ballarat War'.[22]

Given the circumstances, it is not difficult to understand why Burnette, an American, would have fired as he did. When describing the battle for the Eureka Stockade, American author Jay Monaghan alluded to Lexington and Concord, two iconic clashes of the American War of Independence, in which militia fought British Redcoats.[23] He strikes a chord here. There is something very compelling about the image of American miners, part of the Eureka militia, doing exactly the same as their ancestors had done. Reacting to an intuitive impulse, a cultural imperative perhaps, Burnette took aim at the Redcoats. It is not hard to understand why he would have instinctively fired at that instant.

Charles Ferguson. American Charles Ferguson was a member of the Independent Californian Rangers Revolver Brigade. He fought to defend the stockade and was captured following its fall. Ferguson, along with all the other Americans apart from John Joseph, was released from custody a short time later. (Ferguson, *The Experiences of a Forty Niner during a Third of a Century in the Gold Fields.*)

Ferguson claimed that the officer Burnette shot was Wise. Neill described the opening volley from the insurgents as occurring within a short distance from the stockade, and also claimed that Wise fell mortally wounded.[24] Despite these claims, this could not have been the case.

Not surprisingly, given the passage of time and subsequent confusion of the events that followed, both were mistaken in their recollections of when Wise fell. It has been accepted by many in the retelling of the Eureka story that Wise was among the first soldiers to fall. This has been presented over the years as fact, and provided much inspiration for romantic interpretations of the battle at Eureka, both written and visual. Unfortunately, a careful analysis of the evidence suggests that Wise was not among the first soldiers to fall. It is worth digressing from our narrative of the battle for a moment to investigate just when he became a casualty.

Shanahan, who had a tent inside the stockade, had not long gone to bed after a busy night when the first shots were fired. His wife pulled him out of bed and told him to take his gun. When he went out, he saw Wise exhorting his troops.[25] If Shanahan saw Wise after the first shots of the battle, and identified him correctly, Wise could not have fallen mortally wounded as a result of the insurgents' first volley. Other witnesses support a later time, rather than earlier in the battle. Webster recalled seeing Wise get on top of the stockade, and 'immediately afterwards [he] was shot down'.[26] Hegarty recalled that as he 'went into the stockade I saw several of the men down; I saw Captain Wise down',[27] implying that Wise fell at the palisade, or close to it.

O'Keefe recalled that 'I was coming up, [about six paces from the stockade] both me and Captain Wise. Captain Wise fell'.[28] Huyghue wrote that 'Captain Wise on surmounting the pickets fell, shot in the knee'.[29] Despite Huyghue not being an eyewitness, his description tallies with the stories of those who were there. Wise was actually shot twice, the first wound in one leg, the second, mortal, wound through both legs. Webster, Hegarty, O'Keefe, and Shanahan all place him close to the stockade when he received both wounds. Huyghue's account also supports the recollections of those who actually saw him fall. It seems clear that Wise fell close to, or at, the stockade fence, not 150 yards from it as Ferguson claimed, or in the first volley fired by the insurgents as Neill claimed. Who, then, did Burnette shoot?

One other officer, Lieutenant Paul of the 12th, was officially listed as wounded during the battle for the stockade. He could not have been the soldier shot by Burnette, as he was reported by others to have been hit when the infantry closed up to the stockade.[30] There was, however, another officer mentioned as being wounded at Eureka, who did not appear on the official casualty lists. This was Captain George Richard Littlehales of the 12th. Neill stated an officer of the 12th, whose name he could not remember, was hit by the first volley from the insurgents.[31] Littlehales was an officer in the 12th, and died in the government camp on 12 February 1855. His death notice in the *Argus* stated that he died of dysentery and colonial fever, and was buried with full military honours.[32]

O'Brien states that a Captain Little was shot during the fighting for the stockade.[33] Could O'Brien's Little be Littlehales, who was weakened sufficiently by his wound to render him susceptible to dysentery and colonial fever? There is no conclusive evidence. Littlehales was not recorded as having any command responsibilities on the day, nor was he mentioned by any eyewitnesses at Eureka, unless O'Brien's oblique reference is accepted. It is, however, tempting to consider the prospect, even if only speculatively, that Littlehales was O'Brien's Little, whose fate was to be an unsung victim of the battle for the Eureka Stockade.

It is not possible to confirm whom Burnette shot. It may have been a common soldier, perhaps Roney, who did fall, shot through the head, very early in the battle. Burnette certainly fired, and the rifle he carried could certainly have delivered such a shot. Whether this was the first shot fired in anger, also cannot be definitely answered. There may have been several 'first shots' within seconds of each other, the witnesses for each unaware at the time of the other shots, or more likely forgot those competitors to their own claims in the years between the battle and their retelling of the story.

Hackett claimed the first shot was aimed at him.[34] Huyghue, in his second hand account of the battle, stated that the first shot came from the southwest corner of the stockade, and passed over the heads of the mounted troopers.[35] Richards was with Webster and Thomas on the right of the infantry line, and therefore near the southwest corner of the stockade. He gave a vivid account of what he claimed was the first shot, describing it as passing just over Thomas' and his heads.[36] Webster recalled that the same shot 'whistled close by Captain Thomas's head and mine'.[37]

In a prominently published letter to the *Age*, Lalor was adamant that the soldiers fired first, and that the insurgents did not fire at all before this happened. He stated quite clearly that, without warning or provocation from the miners, 'almost immediately, the military poured in one or two volleys of musketry, which as a plain intimation that we must sell our lives as dearly as we could'.[38] Contrary to this claim, it is certain that the first shot came from within the stockade.

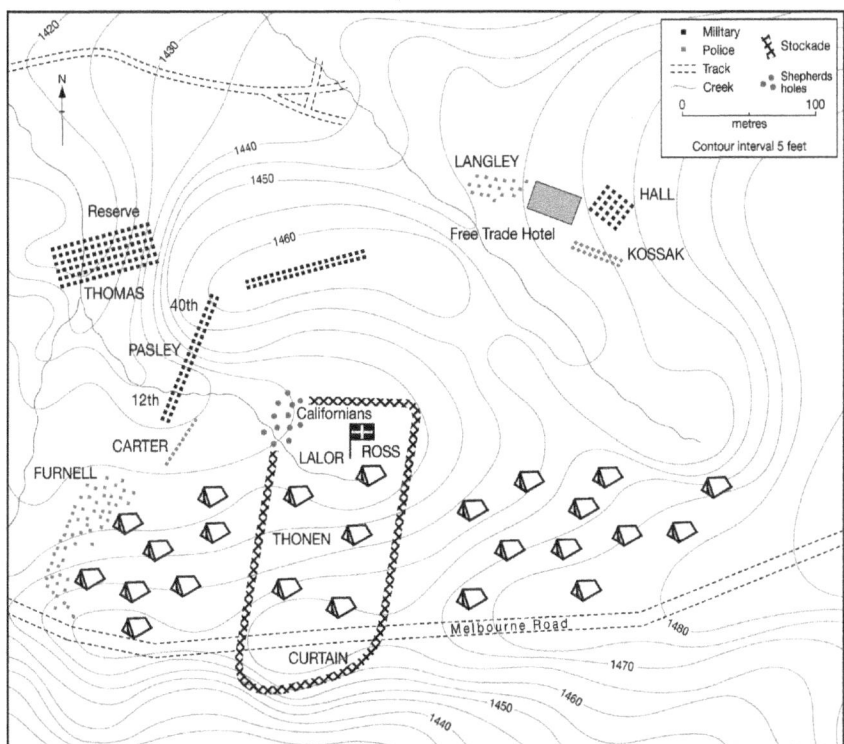

The Battle, phase one: Fire-fight and advance to contact. Heavy firing continued for about ten minutes between the soldiers and the stockade's defenders. During this time some of the 40th wavered and had to be rallied and the police worked their way around the left flank of the stockade.

A note on the tactical maps

Although the exact shape and size of the stockade is not certain, most descriptions by those who saw it indicate its shape was roughly rectangular and that it was something like 150 yards (120 m) in length and maybe half that in width. Likewise the exact locations of the infantry and police attacking, or all of those defending the stockade is not known for certain. These maps, however, depict the action as closely as the evidence for the battle allows us to.

Lalor is mistaken at best, dissembling at worst, and is directly contradicted by numerous accounts from others who were present. In his official report on the action, Thomas made the point that the 'rebels' had fired upon his force, 'without word or challenge on their part'. He was adamant that his men did not fire first, and that a shot, then a volley, had come from the insurgents without warning.[39] Neill, as mentioned earlier, claimed the same. There were also the warning shots heard during the march to the stockade, which definitely were not fired by Thomas' men.

Richards stated that the troops received the order to commence firing after the shot had passed over his and Thomas' head.[40] Hegarty reported first two or three shots, then a volley, from the stockade. His testimony is interesting, as it does imply that there may have been more than one *first shot*.[41] Allen, despite being so deaf as to mishear many of the questions directed to him during the State Trials, gave testimony that while hiding in his tent inside the stockade, he heard several shots go off. He testified that he heard 'pop, pop, pop', followed by a volley that he thought came from the soldiers.[42] Given the obvious caveat applying to testimony from a witness who did not see the action first hand, and had hearing difficulties, Allen's recollections indicate that there were multiple single shots fired before the first volley.

As with almost everything to do with the battle for the Eureka Stockade, there are contradictory opinions about how, when, and what, actually happened. In his deposition to the Ballarat Police Court, Constable William Thompson stated perhaps with exaggeration, that the 'rioters immediately fired … hundreds of shots were fired from the stockade'. He then contradicted himself with a statement that the 'military fired before the rioters opened their fire', and then most confusingly contradicted himself again, by adding that the 'first fire was fired from the stockade'.[43] No reason was given for the discrepancy in Thompson's account, which was officially noted by the court.

Stockade defender Lynch gave a completely different version of how the firing began to Lalor's. Lynch recalled that the first shot, which he says was a single shot and did come from the stockade, was responded to with an 'instantaneous fusillade of musketry' from the soldiers.[44] He made no mention of a volley or anything like it from the stockade before the soldiers fired theirs. Eureka pikeman Miller stated that a man named Henry de Longville fired the first shot, but did not make it clear if this was one of the earlier warning shots, or the first shot in anger.[45]

Charles Knight, writing in the *Mount Alexander Times*, claimed that the soldiers fired the first volley over the heads of the insurgents, which provoked a reply from them.[46] A report by the Ballarat correspondent of the *Argus* on 5 December claimed that a 'demand was made for the insurgents to lay down their arms, or a bugle call at 100 yards. When this was either refused or ignored the soldiers fired two rounds of blank cartridges - the Diggers returned the fire'. While of interest for their unique perspective on how the firing began, the accounts of Knight and the *Argus* correspondent lack credibility. Neither was actually present in the stockade during the battle, nor witnessed it from afar. Knight's account of shots being fired over the heads of the insurgents had its genesis only in newspaper reports of the time, such as those in the *Argus*.[47] The *Argus* correspondent's claim of blanks being fired prior to the engagement appears to be journalistic licence, and is not supported by any other accounts of the battle from either side.

Carboni, who was sleeping in his tent not far from the stockade, heard a discharge of musketry, then a bugle call. This was followed by the command 'forward', and another discharge of musketry. His recollection of a bugle call is of great importance in helping us to determine who fired first. In the British army of the 1850s, the bugle transmitted orders in the field, especially for troops deployed into open or skirmishing order.

It would have been unthinkable, unless in circumstances of undue duress, that British infantry would fire without orders. Such insubordinate initiative by the common soldier of 1854 could see his back flogged raw. When Thomas had earlier told his men to wait for the sound of the bugle before they fired, he was reminding them of what was expected of them.

To claim that the order might have been whispered, spoken, or even shouted is unrealistic. No one present that morning reported hearing any verbal order given to the soldiers to open fire. That would have been a problematic method of transmitting orders. Not all of Thomas' men would necessarily hear it, even if shouted, and the potential for confusion would have been great, as firing suddenly erupted from soldiers on only one part of the field. Confusion is anathema to a commander in battle, and Thomas would not have done anything to encourage it, especially as the success of the operation depended on quick and decisive action.

It could, of course, be argued that the representatives of the military and goldfield authorities, such as Thomas, Hackett, Pasley, Hegarty, Richards and Neill, would claim that they did not fire first. By doing so, they would shift any blame for what followed away from themselves and the government. However, their accounts concur with those of Ferguson, Shanahan, and Sutherland, none of whom had a vested interest in trying to shift blame on to the insurgents. Carboni's account of when he heard the bugle certainly supports the contention that the soldiers waited for the order to fire. It seems certain that Ferguson was relating what actually happened when he wrote that it 'was said by many that the soldiers fired the first shot, but this was not true, as is well known by many'.[48]

Whichever individual or individuals fired the first shot or shots in anger, that action released whatever inhibitions remained within the men who crouched inside the stockade. Numerous shotguns, a few rifles, and maybe even a musket or two, lifted. Then a ripple of flame and smoke erupted from behind the stockade and among the shepherd holes, as the insurgents fired a volley into the advancing Redcoats. The effect of that volley was significant.

The effect on the soldiers of the insurgents' opening volley impressed itself on Neill. He recalled four men of the 12th and four of the 40th, mistakenly including Wise, falling. One of the first casualties was Roney,[49] who was shot through the head and died on the spot. W. Bourke, who lived 250 yards (225 metres) from the stockade, saw a soldier fall and be carried away by two others, who then returned to the battle.[50] This may have been Roney.

Other men went down. Privates Felix Boyle, William Juniper, and possibly John Wall, all fell. Boyle, a 32 year old veteran of the Sikh Wars, was hit square in the face, a ball entering his head near his nose. Lynott recalled seeing a soldier lying under a windlass after the first volley, and that the man died the next day.[51] This was probably Boyle, who did die the next day. Juniper had a bone in his leg smashed so badly by a ball that it caused a compound fracture.[52] According to his military record, Wall died from pike wounds received during the battle, but on his death certificate he was listed as 'shot and spiked'.[53] The records are silent on when Wall received his gunshot wound. Perhaps he was hit by the first volley, but not injured sufficiently to prevent him carrying on, and eventually met his demise on the point of a pike inside the stockade.

After having taken fire, and with men down, Thomas rode across and ordered his bugler to sound 'Commence Fire'.[54]

With a throaty tympany and belch of white smoke, several score military muskets sent their 0.753-inch calibre soft lead balls whining across the space between their firing line and the stockade's defenders. The slow velocity of the musket balls created a distinct sound as they passed through the air. To the firer, the sound is something like a whine that forms a descending crescendo as the ball flies away. To those at the receiving end, the sound could be like the 'whizzing of mosquitoes'. Chips splintered from the stockade's slabs as the balls hit.[55] No doubt some insurgents went down. Shanahan says he saw nine lying dead as a result of the soldiers' first volley.[56]

With their initial volley fired, Thomas' skirmishers would have employed steady, tried and true tactics, certainly not the totally random and wild free for all described in one account.[57] When the order came for the skirmishers to commence fire, the system used was known as a 'chain'. The man on the right of each four-man section stepped forward three paces and fired, then stepped back and re-loaded. The next three men did the same individually, when the process started again. This system ensured that fire was always being delivered against the enemy, and that the line always presented a partially moving, extended order, target, and was thus much more difficult to hit.

Agnes Greig, who saw the attack on the stockade, mentions seeing soldiers kneeling on the ground as they fired, which was standard for skirmishers.[58] The chain was not intended to hold ground, and not necessarily to advance. When the order came to move forward, the skirmishers moved and fired in bounds, with each man running forward in a controlled manner, firing in turn, and holding the new ground. Working in this way maintained a steady rate of advance, as well as ensuring that a quarter of the line was firing as the other three quarters were preparing to do so, an important consideration in an era of single shot weapons.

Unlike the beginning of the nineteenth century, when only specialised troops were trained as skirmishers, by 1854 all British soldiers practiced skirmishing as one of their tactical drills. This was because the skills learned were essential as the British Empire grew, and its soldiers found themselves matched against an increasingly unconventional array of enemies in exotic locales around the globe. Skirmishers did not necessarily engage their enemies in hand-to-hand combat, but would do so if compelled to, in places such as wild forests and mountains, where traditional shoulder-to-shoulder tactics were useless.

The military consensus of the day, however, was that when it came to serious bayonet work, closer formations were preferable. Consequently, bayonets were not fixed when troops were skirmishing. If an assault looked to be in the making, or the tactical situation appeared unfavourable, the skirmish line supports were brought up. This is exactly what Thomas did at Eureka, as the situation developed into a serious confrontation.

As the soldiers fired into the stockade, the insurgents returned what Lynch, Ferguson, Thomas and Pasley described as a 'sharp' fire. About this time, Carboni emerged from his tent outside the stockade, and claimed to see about a dozen soldiers on the ground, ample evidence of the effectiveness of the defender's fire.[59]

One part of Thomas' force found itself in a particularly difficult position. The foot police had advanced more quickly than the soldiers, moving ahead before firing had commenced, to a position somewhere between the skirmish line and the stockade. Now that shooting had started, they found themselves between the fire of the soldiers and the insurgents.[60] With bullets and shot coming at them from the front, and military musket balls from the rear, they had no choice but to fall back to their original position on the right of the 12th, where they remained until the moment came to enter the stockade.

To understand the practical dynamics of the battle, and the nature of what the men on both sides faced that December morning in 1854, it is important for us to understand the capabilities and effectiveness of the weapons being fired at and from the stockade. For trained troops, the rate of fire for single shot muzzle-loading smooth bore weapons was normally no more than three times a minute. In circumstances of *extremis*, the rate could be increased to four of five times a minute. However, such rapidity of fire came at the cost of appallingly bad accuracy.

With only a few exceptions, the firearms employed by both sides were notoriously inaccurate smooth bore weapons. Most of the long firearms carried by the insurgents were shotguns, good at short range, but severely limited beyond 50 yards (45 metres) when firing shot. The soldiers' smooth bore muskets were little better, but at least had a chance to hit targets out to about 100 yards (90 metres). The opening volley by the soldiers was fired at about this distance, or perhaps a fraction more.

This is significant, as it would be highly unlikely that more than two to five percent of those shots scored hits on the defenders, or even the stockade itself at that range. This may seem to be an absurdly low probability, but is amply supported by extensive analysis of battlefield casualties inflicted by firearms during the smooth bore era.[61] There were 176 soldiers at Eureka. Thomas estimated that the exchange of fire continued for about ten minutes. Even if this estimate was incorrect by a few minutes, with the number of soldiers present and the rate of fire they could achieve, many hundreds of musket balls would have been fired at the stockade.[62] A lower, yet still imposing, return fire came from the defenders. This would have been an impressive display. The continual heavy reports of muskets and shotguns, muzzle flashes, and clouds of billowing smoke, establishing without doubt the significant nature of the conflict.

The soft lead balls could produce horrific injuries, yet if vital organs were missed, a wound did not necessarily cause immediate disablement of the victim.[63] The musket and pistol balls fired at Eureka behaved in an entirely different manner to modern rounds. Because they travelled at the relatively slow velocity of 600 feet (180 metres) per second,[64] the lead balls tended to burrow into flesh, leaving a passage of uniform diameter, the same as the calibre of the ball. Only the tissue in the immediate vicinity of this wound passage would be damaged. The result was that if vital organs or bones were not hit, a victim could, if fortunate, remain able to move and fight.

An illustration is Michael Canny, who was in the stockade. He described the wound he received in some detail, as 'a bullet pierced my right arm, went in my side, and out under my breast bone. It did not hurt, but the blood spurted out, and scared me'. He explained that he was able to leap over the stockade fence and ran away. Canny's brother, Patrick, was not so fortunate, being hit in the shin and having the bone 'splintered'. [65] Michael was very lucky, but his experience was not unique, and clearly shows that a wound inflicted by a musket ball might not immediately incapacitate a victim.

It was quite common in battles of the black-powder era that victims of gunshots received more than one hit and still remained active, requiring additional shots before succumbing. The description earlier of a British officer firing six shots into an attacker without immediately killing the man is a case in point. The wounds suffered by Wise are another example. His initial wound was in the leg, and knocked him down, but he regained his feet. When he was hit a second time, bones were shattered, and the wounds were much more serious. Unable to regain his feet, Wise nevertheless remained conscious, and encouraged his men onward.

The multiple gunshot wounds suffered by many insurgents at Eureka have provided the basis for claims that the soldiers inflicted gratuitous violence. Such claims rest on very shaky ground. The soldiers who stormed the stockade were armed with black-powder muskets, and the police armed with black-powder pistols, shotguns and carbines. Many examples show that a single wound from such weapons might not kill, or even incapacitate, an enemy. Multiple gunshot wounds in the bodies of insurgents at Eureka should therefore not be surprising. The soldiers of the 12th and 40th, and the police, would have been well aware that it might require more than one shot to kill or incapacitate. To accuse them of inflicting deliberately malicious and excessive violence on the Eureka insurgents ignores the nature and (limited) effectiveness of the weapons of the era.[66]

What did the soldiers face as they advanced against the stockade? One of the deeply entrenched myths of Eureka is that, because the insurgents were supposedly so poorly armed and caught unawares, there was very little effective fire from them. One source goes so far as to describe their fire as an, 'intermittent splutter'.[67] Despite such casual dismissals of their efforts, the reality was certainly impressive enough for the men who were on the receiving end of it.

Huyghue reported that, in a conversation with a policeman after the fall of the stockade, the constable described the number of bullets being fired to and from the stockade as 'flying about thick as hail'.[68] Pasley described the fire from the insurgents as '[s]harp and sustained'.[69] Thomas referred to the fire coming at his men as 'rather sharp, and well directed'.[70] Lynch referred to some 'sharp shooting' when the firing began.[71] Ferguson remembered the insurgents giving back as good as they got, describing their fire as 'with like effect, as deadly as theirs'.[72] When Thomas and Pasley referred to the fire coming from the stockade as 'well directed' and 'sustained', they were describing what they were actually experiencing. Both men were military professionals. Thomas, in particular, had seen war at its most intense.[73] Such men would not be given to flippant hyperbole.

Shocked by the intensity of the fire coming from the stockade, some of the men of the 40th began to fall back out of line. As the bullets from the insurgents cut through their ranks, and some found their marks, there was a momentary failure of morale. Carboni recalls seeing the soldiers' line 'swerve from its ground'.[74] Lynch mentions that 'the advance of the infantry was arrested for the moment'.[75] Huyghue commented that the severity of the insurgents' fire 'caused the Queen's infantry to waver and many of them held back'.[76] How could such a thing have occurred?

Some of the Redcoats were very young. Huyghue had earlier referred to them as '[s]triplings', with the added flourish of 'half weaned cubs of the Lion Mother'.[77] He supplemented this image of inexperienced youngsters suffering at the hands of savage fire when he described many of them as 'raw recruits who were mere boys in fact'.[78] His apology for the soldiers' subsequent failure wavering under fire does not hold up to close scrutiny.

When the ages of the soldiers who are known to have been at Eureka are considered, a very different picture emerges to the one Huyghue paints. The average age of men of the 12th known to have been at Eureka was 21.7 years, with a significant number of 18 year olds in the ranks. This is understandable, as the 12th had arrived in Australia only in November 1854, direct from England, where many of the men had been recruited in 1853. This might justify their description as 'half weaned cubs'. On the other hand, soldiers of the 40th who were at Eureka, and whose ages are known, had an average age of 28.2, with several men older than 30 in the ranks.[79] Huyghue's excuse of the youth of the soldiers clearly fails, because it was the 40th, which included the older and more experienced troops, that gave ground, not the 12th. When these troops, the best Thomas had available, wavered, it was stark evidence of the effectiveness of the defenders' fire.

When the men of the 40th wavered, the seed was sown for what could have become a very serious situation. Panic is infectious, and can spread rapidly if not nipped in the bud immediately. Thomas' whole plan of attack would have hung in the balance. At that crucial moment the bulwark of the British army, the sergeant, intervened.

It is a truism that the backbone of any professional army is its senior NCOs. This was especially so in the British army, and it was proven once again on the bullet swept ground in front of the Eureka Stockade that December morning. Carboni, watching from outside the stockade, says he heard Sergeant Harris of the 40th shout 'Forward!'[80] It is tempting to speculate what else the good sergeant may have bellowed.

Wise, unwounded at that time, and seeing his men losing heart, also called out his encouragement. Shanahan said he thought he heard the officer call out 'Fortieth are you going to retreat?'[81] Carboni also recalled a boy bugler bravely standing his ground and the men, rallied by the efforts of Harris and Wise, formed up on the boy's right[82]. About this time, Thomas, realising that he must intervene and restore his men's spirit, dismounted and took direct command of the troops in front of the stockade.[83] His action highlighted how serious he considered the situation had become. It was quite possible for him to have been hit by a shot from the stockade, and his death or wounding could have led to an unravelling of the attack.

Thomas would have been fully aware of that, but the situation demanded his presence. Such actions were not untypical of good officers of the era. When warned not to expose himself to enemy fire at a moment of crisis during the American Civil War battle of Gettysburg, Major General Winfield Scott Hancock replied that there 'are times when a corps commander's life does not count'.[84] In much the same mind, Thomas stepped forward to lead and encourage his men.

Fire that can cause a line of experienced British regular infantry to waver is not to be dismissed as an intermittent splutter. It was also not the fire of men who had been caught napping, or were 'saturated with drink', as one of the more persistent myths of Eureka would have it.[85] It was fire that was indicative of a determined and competent defence that implied at least some, and perhaps more than a few, of the defenders were quite capable of delivering 'sharp and well directed fire'. There are two ways this could be interpreted. The first is that the fire was well aimed, a particularly pertinent point when firing from higher to lower ground, the second that the fire was controlled and delivered in a coordinated manner. It seems that, for a while at least, both types of fire were coming from the stockade. Even so, its effectiveness would vary.

Regardless of how well aimed the insurgents' shots were, there were not many weapons inside the stockade capable of reaching out to the range of the soldiers when the exchange of fire began. Whatever effective fire came from the insurgents in the opening stages of the battle would have been from those armed with rifles and muskets. Burnette and his rifle are a good example of this. Some insurgents would have also been firing larger bullets from their shotguns. Even so, well-aimed individual shots delivered randomly would not have been as noticeably effective as well directed fire; fire that is concentrated and delivered in unison, or near enough to it. Such fire has the best chance of producing traumatic effects on its target. Who within the stockade had had the skills to compensate for firing at their foes from higher to lower ground, as well as to control and direct their fire?

One group that comes immediately to mind is the Californians of the Independent California Rangers Revolver Brigade. While most of the Rangers had gone out to Warrenheip with McGill, to intercept the reinforcements coming from Melbourne, some 20 or 30 of them remained in the stockade. Ferguson refers to many of the men who joined the California Rangers as being veteran 'Mexican soldiers of 46-48'.[86] Allan also mentioned the presence among the American insurgents of 'old soldiers in the Mexican war'.[87] This would not have been surprising, as that war was fought only a few years before Eureka.

The presence of Mexican-American War veterans inside the stockade is also mentioned by other sources. Carboni implies that Johann Hafele, a German blacksmith, was one, describing him as 'praising the while his past valour in the wars of Mexico'.[88] Carboni goes on to further imply some sort of American connection, by claiming that while Hafele was making pikes at the forge he was 'making money as fast as any Yankee is apt to on such occasions'.[89]

It was quite possible that Hafele served in the Mexican-American War. The United States army was a haven for immigrants, with Germans and Irish over-represented in the ranks. During the 1830s it was estimated that two thirds of the common soldiers in the United States army were foreign born, with Germans, Irish and British making up the majority of recruits. The situation had not changed much by the outbreak of the Mexican-American War in 1846.[90] If a significant number of the Californians, Hafele, and perhaps others unnamed inside the stockade, were veterans of that war, they would have constituted a group of men within the stockade who were used to accepting direction in battle, and capable of exercising the fire discipline necessary to produce the sharp and well directed fire described.

There were others in the stockade who might also have used firearms in conflict. An Italian named Oravalano, who was killed during the battle, was said to have served with the Italian revolutionary Garibaldi in 1848, implying he had experienced combat.[91] However, apart from Oravalano, a search of the names of the men known to be inside the stockade, other than the Americans, does not reveal any with recorded prior military service.[92] This does not mean that there were none, but references to a lack of military acumen among the insurgents drilling at Eureka imply that there were very few. With no other contenders, this leaves the Californian-American contingent as the only identifiable group of defenders who might have possessed military skills and experience.

We already know that the Independent California Rangers Revolver Brigade rushed to the wall in the opening stages of the battle, and Ferguson claimed that Burnette fired the shot that most probably ignited the savage exchange of fire between soldiers and insurgents. According to Carboni, the Californians were posted in the lower part of the stockade.[93] They were behind the barricade there, and also occupying shallow shepherd holes that had been turned into rifle pits. A square framework of logs, high enough to pack dug-up earth around them, normally defined the edges of shepherd holes. Sometimes a windlass would be erected above them.[94] Such positions provided quite effective shelter from which men could fire at their enemies. It was the fate of the Californians that these holes, and the slab barricade at the northwest point of the stockade, lay in the direct line of the main military assault.

The Californians were better armed than the average insurgent. As their Brigade's name implied, most of them were armed with revolvers, probably Colts of various calibres. The incident with Burnette and the effectiveness of the fire that was delivered at ranges beyond that of revolvers indicates the significant presence of long arms, such as rifles or double-barrelled shotguns, as carried by Lynch who, though not a Californian, was assigned to them.[95] They also possessed adequate ammunition, or at least enough to sustain their fire for something like ten minutes. Even though any of the Californians' military skills might have been somewhat rusty by 1854, old soldiers never really forget what they have learned, especially the lessons taught by war. These skills were augmented by some effective and energetic leadership.

The Californians at Eureka. There were twenty to thirty Californians inside the stockade during the battle. They contributed significantly to the defence of the stockade and, when all hope was lost, a number of them counterattacked the victorious soldiers in an act of desperate courage and bravado. (Original panting by Gregory Blake)

An American mentioned by Carboni, who commanded the 'rifle pit men', by whom he must have meant the Californians, played a significant role in the defence of the stockade. Despite being wounded in the thigh, he 'fought like a tiger'.[96] If it was not this man who controlled the fire that so discomforted the men of the 40th, who was it? It was almost certainly not Lalor who, despite boldly exposing himself to the soldiers' fire and shouting orders, had very little real impact on the battle as it unfolded. The Italian Oravalano, if he actually was a Garibaldist, may have directed the fire. However, he is only mentioned in a newspaper report. If he had been noticeably directing the defence of the stockade, surely someone present would have said so.

It may have been Ross or Thonen, but this is unlikely, as neither had military experience. The American mentioned by Carboni remained in close contact with Ross throughout the battle, perhaps inspiring Ross's men also to fire in a coordinated manner. This American, who was never named, appears to have been a crucial factor in the opening moments of the defence. Armed with revolvers and long arms, and drawing on their military experience, the Californians, under his direction, could have been capable of delivering the weight of fire that so impressed the stockade's attackers.

The casualties among the attackers are further evidence of the effectiveness of the insurgents' fire. Although it is not possible to state exactly when each of the military casualties occurred, it is possible to make an informed estimate for them as a group. Of 19 military casualties suffered at Eureka, 14 were definitely due to gunshot wounds. As mentioned earlier, Neill reported Wall being hit by a shot outside the stockade, before he suffered a pike wound. If Wall is included as a gunshot casualty, the percentage of soldiers killed and wounded by gunshot increases.

There is no reason to question Carboni's recollection of about twelve soldiers on the ground behind the skirmish line early in the battle. These twelve casualties represent 86 percent of the total known gunshot wounds, or 63 percent of the total casualties suffered by the Redcoats during the battle. According to Carboni, these casualties all occurred early in the opening stages of the battle, before the infantry reached the palisade. Knowing that most shots fired in battle actually miss their intended target, this emphasises again the effective and heavy fire from the stockade at that time.[97] Only a collective competence, not random individual efforts, could produce such fire. The only group inside the stockade that morning that could have done so were the Californians.

In his account of Eureka, Carboni made little mention of the Californians, except to say that they were there. His oversight has been maintained and refined by the many Eureka storytellers who followed, with the result that the very significant contribution they made to the battle at Eureka has been almost completely ignored. The reasons for this reveal much about the political situation at the time, and the contest for ownership of the Eureka myth since.

When James Madison Tarleton, the United States Consul to Victoria, and a personal friend of United States President Franklin Pierce, declared that 'there are not any Americans engaged in the affair',[98] he was indulging in some very wishful thinking. Hotham soon set the consul right, by referring to an American, by whom he meant McGill, as being 'their most active leader'.[99]

However, Hotham was determined not to pursue the American connection with what had happened at Eureka, for several reasons. The American business community in Victoria was prominent and influential, and not making enemies of them made good sense. It was also wise not to cause diplomatic problems with the United States by actively pursuing and punishing American citizens when Britain was involved in a war with Russia in the Crimea, and a potential revolution was brewing in Victoria.

Hotham therefore embarked on a deliberate policy of minimising the repercussions for those insurgents who were American. His largesse towards them went so far as to excuse even McGill, the leader of the California Rangers, who remained a fugitive for a while, then, following the intervention of Train, was quietly forgotten. In like manner, the authorities turned a blind eye to the involvement of all other Americans in the insurgency, with the exception of the African-American, Joseph.[100] This caused much bitterness among other miners at the time.[101]

Americans during the early 1850s did not enjoy uncritical acceptance on the Victorian goldfields, or in the wider community. The attitude of the colonial Anglo-centric establishment to all things American was one of consistent low-level ambivalence, which at times verged on antagonism. Americans were admired for their energy and business sense on the one hand, but patronised and shunned on the other. American democracy and the rough justice meted out to criminals in California were consistently criticised, and Americans were occasionally lampooned individually as showy and somewhat pompous characters.

Charles 'Henry' Ross, the Canadian 'Captain' of a rifle company of diggers at Eureka. Ross was a leading personality amongst the insurgent miners who refused to leave the Stockade when he was told before the battle that the military were approaching. Ross was Mortally wounded beneath the Southern Cross flag during the battle. *Photo courtesy if Ian MacFarlane.*

Examples of such attitudes are numerous. The hangings in San Francisco of notorious 'Australian' criminals James Stuart and Samuel Whittaker, and the Scot Robert McKenzie, excited outpourings of sanctimonious moral outrage in the Australian press. On the arrival of the first ship carrying Californian gold seekers to Australia in 1851, the editor of rhe *Empire* demanded that 'no door be opened to receive the blood stained wretches'. Brash, bragging, and flashy American stereotypes were ridiculed in song and theatrical skits. Americans and their ways were gently mocked in satirical magazines.[102] The existence of slavery in the United States, more often than not portrayed in its most lurid form, was used as the proof that exposed the hypocrisy of Americans championing liberty and equality. The widespread popularity in Australia of such books as *Uncle Tom's Cabin* fuelled such prejudices.[103]

Pierson, an American miner whose tent was near Bakery Hill, mentioned that there was a noticeable prejudice against Americans by others on the diggings, born of jealousy.[104] Daniel Selman recalled that he had to push his possessions to the diggings in a wheelbarrow because, as an American, he was denied assistance to get there by any other means.[105] Although as law abiding and hard working as any other part of the miner population, Americans were often portrayed by the press and judiciary as lawless and violent personalities.

The case of an American miner, John Clarke, who in a drunken and contrary mood fired his revolver during a card game inside the Albion Hotel at Ballarat, mortally wounding its proprietor, was used as an example of the inherently lawless and violent nature of Americans.[106] The judiciary's attitude reflected these perceptions. Judge Redmond Barry, when sentencing an African-American hotel cook to ten years hard labour for using his knife to attack an intruder, chastised the unfortunate man's choice of weapon as 'un English', and his method of responding to the unwelcome trespass as 'unknown among British subjects'. Another judge advised two Americans who had pulled their knives on each other, but not drawn blood, that their actions did not meet with the approval of English law.[107]

Prominent Legislative Council member John Pascoe Fawkner, a consistent critic of Americans, characterised as 'murderous' the Bowie knife, an iconic weapon carried by many American miners.[108] Concerns for the supposed clandestine republican game being played by Americans, as they plotted to Americanise Victoria, exercised the imaginations of government officials such as Rede.

The basis for this bias was a fear among British traditionalists of the challenge American ideas posed to the established political and cultural order. In a denouncement of how Americans felt it their right to become involved in government, the *Argus* warned those locals tempted to emulate the American model that such notions of asserting one's citizenship by direct action were an abrogation of British citizenship.[109] Thus, according to the *Argus*, the very nature of British citizenship was predicated on the proposition that the common people did not have the right to indulge themselves in directly determining who governed them. Such was the understanding of how a respectable British society should function.

By championing democracy, Americans fostered a social model directly threatening to that ideal. In 1855, the American poet Oliver Wendell Holmes observed that 'the English see us through the medium of insular prejudice, the conceit of their little island'.[110] In the failure to acknowledge and commemorate the role played by the Americans at Eureka, Holmes' words ring all too true.[111]

In the climate of ambivalence towards Americans that prevailed in colonial Victoria, it is understandable that Hotham's overt favouritism following the traumatic events of Eureka excited animosity. This was the genesis for at first the casual dismissal, then the eventual deliberate denigration, of the contribution made to the defence of the stockade by those Americans, including many Californians, who were present that day.

In some cases, such as the criticism of McGill and most of his Independent California Rangers Revolver Brigade for being absent from the stockade when it was attacked, the vilification of Americans at Eureka assumed all the unsavoury and paranoid aspects of a classic conspiracy theory. Carboni, in one of his frequent digressions into the obscure, equates the role of the Americans to that of Lot's Wife, perhaps meaning that the Americans stood and watched as a holocaust consumed their peers.[112] Lynch, bitter over the defeat, cast aspersions on McGill's role in the debacle.[113] John Ashburner recalled that some had accused McGill of being a traitor.[114]

Other, more scholarly, accounts persist in the slander that McGill deliberately deserted the stockade prior to the attack.[115] Such accusations are facile. To presume that McGill, Nelson and the several hundred insurgents with them were all participants in some grand conspiratorial plot, and all kept that secret for the remainder of their days, is preposterous. This is especially so as there were 20 to 30 Californians inside the stockade. These men would presumably have friends and acquaintances among the men who accompanied McGill and Nelson to Warrenheip. To suggest that the men at Warrenheip would have remained silent if their friends had been sacrificed in such a manner is absurd.

Another reason the men at Warrenheip did not intervene at Eureka is the distance they were from the stockade. Depending on exactly where they were, they could have been from three to five miles (five to eight kilometres) away. Given the undulating and wooded nature of the terrain, it is not certain that McGill and Nelson could hear the shooting at all, and even if they did, that they would understand what it meant. Even if they heard something and recognised it for what it was, it was not possible for their men, most of whom were on foot, to cover the distance from Warrenheip to Eureka in the 20 minutes it took the attackers to overrun the stockade.

The lack of records asserting that the men at Warrenheip heard anything explains why they did not move towards the stockade during the fighting. Suggestions that they did not do so because they deliberately chose to betray the Eureka insurgents are nonsense. Other spurious claims bedevilled McGill's reputation following Eureka. Later allegations by his wife that the United States Consul instructed the Americans to remove themselves from the stockade prior to the attack can be understood as an attempt by a wife to rescue her husband's reputation. No proof was ever offered that such an order was given. Vern, who had expressed contempt for the Americans after their meeting at the Adelphi Theatre, claimed that McGill had taken an £800 bribe from the government to absent himself from the stockade. No evidence of this supposed bribe has ever surfaced.

McGill, and by association all of the Americans, were made scapegoats for the catastrophe that overwhelmed Eureka's defenders. With the prevailing ethnocentric prejudices and political circumstances, this was easy to do. As a consequence, the contribution of the Americans at Eureka was diminished at the time and since, to a point where it is frequently not even acknowledged.

Sadly, the lack of interest by succeeding generations in the contribution made by the Americans to the defence of the Eureka Stockade has been endemic. W.B. Withers, in his classic *The History of Ballarat*, a source for many who study the events at Eureka, does not mention them. Other authors and historians since have followed his lead, and ignored, or at best acknowledged it only in passing, the American participation in the battle. Given the events that immediately followed Eureka, and the prejudices against Americans and things American found within British colonial society at the time, it is easy to see why this occurred.

As American anthropologist John Greenway astutely observed, the removal of Americans from the story also 'purified the incident for Australians'.[116] His point is that the power of heroic national mythologies stems from their intensely parochial nature; therefore it just would not do to concede the significant involvement of supposed outsiders in the events that created the legend. While a few modern historians, to their credit, concede that the Americans fought well at Eureka, there is never any attempt to investigate just what they did contribute to the battle. This does an immense disservice to their memory. It is surely the time such misconceptions were put to rest.

When the men of the 40th wavered in front of the stockade, for a brief moment the fortunes of the attacking force hung in the balance, but the moment passed. Chastised by their officer and sergeant, the soldiers re-established their line and began advancing again, as each man fired in his turn. The insurgents inside the stockade returned the fire.[117]

Not all of the fire against the soldiers that morning came from the stockade. Lieutenant Thomas Gardyne of the mounted 40th reported seeing shots fired at the skirmishers from tents outside the stockade.[118] Thomas' men themselves did not report these shots, but in the half-light, smoke and tumult of battle, that would be understandable. Perhaps these shots struck home, perhaps not. Either way, they appear not to have influenced the progress of the action. This report does indicate, though, that not all the defenders were behind the slab palisades that morning, and helps to explain why tents in the immediate vicinity of the stockade were set upon with such zeal by soldiers and police immediately after the action.

About ten minutes after the attack commenced, the mounted police troopers began to make their presence felt. They had been sent around to the west and rear of the stockade when it was still relatively dark. This proved to be an astute move. With their depleted numbers, the insurgents had not been able to cover that part of the stockade adequately. As the light improved, this was noticed by the attackers, and was now exploited by the troopers who, in Lynch's words, 'wheeled round, and took us in the rear'.[119]

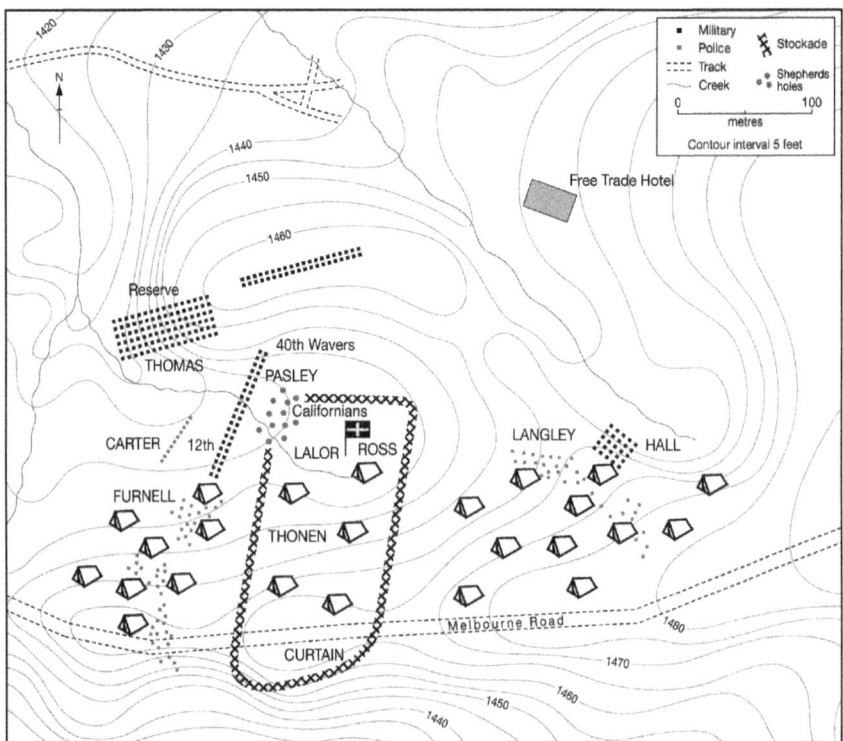

The Battle, phase two: The infantry assault. With the police firing into the flank and rear of the stockade's defenders the infantry were able to close up to the stockade and assault it with fixed bayonets.

Most of the troopers were armed with muzzle-loading carbines and pistols. Some would have had revolvers, and a few had muzzle-loading percussion cap shotguns. One mounted police unit at Eureka was equipped with rifled weapons that day, Polish soldier turned police officer Kossak commanded a unit of troopers armed with rifled breech loading carbines.[120] At least one trooper was armed with what Carboni rather cryptically describes as a 'minie rifle pistol', which was probably a large muzzle-loading cavalry pistol fitted with a detachable carbine stock, a type of weapon then used by British mounted units.[121]

The police pistols and revolvers were useful only at very close range. The carbines they used at Eureka were smooth bore weapons, and suffered all the failings of the larger muskets used by the infantry. Yet, like the muskets, they could deliver a devastating wound if luck favoured the shooter. Despite the best efforts of novelists and screenwriters to persuade us otherwise, hitting targets while firing from horseback is notoriously difficult. It can be therefore assumed that, to deliver some sort of effective fire, which they did, some troopers dismounted to fire while others held their horses.

In addition to their firearms, the mounted police and soldiers carried swords, which they made much use of that morning, especially in the pursuit through the diggings that followed the fall of the stockade.

Some of the police, and perhaps mounted soldiers, who had moved out to the east of the stockade now took post on high ground behind the Free Trade Hotel. From that elevated position, about 200 yards (180 metres) from the stockade, they began to lay down covering fire to support the left of the infantry line. The site, and even the name Free Trade Hotel, are disputed.[122] Whatever the case, there was high ground behind a 'hotel' near the stockade. The distance from the hotel to the stockade is important in giving us a clue who occupied the high ground near it.

A distance of 200 yards is well beyond the effective range of smooth bore weapons, especially the short cavalry carbines many of the mounted police carried. Such a distance is not, however, beyond the range of rifled weapons. If anyone provided fire from the high ground at the rear of the Free Trade Hotel, it must have been Kossak's troopers with their rifled breech loading carbines .

With infantry continuing to exchange fire with them from their front, and the police now firing into their right flank and rear, and dropping fire into the stockade from near the Free Trade Hotel, the moment of crisis for the Eureka insurgents was fast approaching.

Chapter 8
THE REDDENED CLAY OF EUREKA

For about ten minutes there had been a heavy, continuous, fire between the soldiers and the insurgents.[1] Sometime during that exchange, the men of the 40th wavered. To at least some of the soldiers held in reserve, it might have seemed that the attack was in danger of stalling, or even failing. This was the probable explanation for a most curious event that occurred.

As balls and shot flew between the soldiers and insurgents, some men from the reserve ran forward, braving the fire from both their comrades and the insurgents. Reaching the slab fence of the stockade, they jumped over among the defenders. Bourke, who lived about 250 yards (225 metres) from the stockade, described five or six soldiers running ahead of the main body of troops and jumping the barricade.[2] Their incursion was short lived. Almost immediately, insurgents armed with pikes attacked them and chased them out. The muskets of the main body of soldiers were then turned on the pikemen, shooting down five or six.[3] It cannot be confirmed why these soldiers acted in this manner.

Huyghue described this attack as being delivered 'pell mell', implying that it was an unplanned act by the men concerned.[4] He did not mention the intervention of pikemen. There was only one known serious casualty inflicted by a pike during the battle for the stockade. Wall suffered a mortal wound to the lower abdomen. Even though Bourke did not mention any soldier being injured during their mad dash and subsequent retreat,

Wall may have suffered his wound then. That something compelled the handful of Redcoats to flee is indisputable, probably the understandable fear of being impaled by a pike. Whatever their reason for acting as they did, it seems that once inside the stockade, the impetuous soldiers allowed discretion to overcome their valour, and retreated very quickly to the relative safety of their own lines.

This was a curious incident. It would have been out of character for common soldiers to act without orders, a misdemeanour that was ruthlessly discouraged during the era, by the lash if necessary. Such behaviour was not completely impossible, but highly unlikely, and begs the question, just what did happen?

An explanation might be found in the number of unattached officers who accompanied the force attacking the stockade. One of these was a Lieutenant Adams of the 40th, who did not appear on any rolls for the regiment at the time of Eureka, but was mentioned as being at the battle.[5] Then there was Captain Thomas Nelson, also of the 40th. He was a veteran of campaigns in Afghanistan and India, and was listed as being present at the storming of the stockade, but does not seem to have had any official role.[6] O'Brien also claimed that a Captain Little, perhaps Littlehales, was present. Wall was from the 40th, the same regiment as Adams and Nelson. Did one of these officers, or even another not named, frustrated by what the perceived unravelling of the attack, take matters into his own hands, and order some soldiers from his own regiment to rush the stockade?

All through the first ten minutes of firing, the Redcoats of the skirmish line edged forward, halted only momentarily by the brief moment of panic among the 40th. From the protection of their musket-proof slabs, the insurgents continued firing. As they did so, they made sure to keep their heads down as much as possible, prompting Harris to recall later that the defenders of the stockade were 'rather shy, and kept their faces on the side'.[7] Nevertheless, as the soldiers slowly advanced, the fortunes of battle were shifting inexorably against the defenders.

The volume of fire into the stockade increased steadily. In addition to the fire of the skirmishers and the soldiers on Stockyard Hill, the troopers and soldiers at the Free Trade Hotel were firing. Even so, the insurgents, protected by their stockade, were still holding their ground and giving back as good as they got. It would need something more decisive to break their defence. Just such a threat was developing along the stockade's western flank.

When the attacking force had deployed from their march columns in the pre-dawn darkness, 70 or so mounted troopers led by Furnell, with Amos as a guide, worked their way around to the west of the stockade. In his original plan, Thomas had intended Furnell and his troopers to threaten that flank and the rear of the stockade. His foresight was about to pay off.[8]

Despite their lack of numbers, it appears that the insurgents facing the troopers along the western flank of the stockade gave a good account of themselves. Huyghue described how a trooper faced shot as '[t]hick as hail'. The same man had a rein cut by a bullet.[9] White recalled 'many shots being fired at us'.[10] The insurgents facing Furnell's police certainly seemed prepared to do what they could to repel the police to their front.

Despite White's claim they were the targets of 'many shots', and Huyghue's description of a hail of shots, there were no recorded gunshot wounds suffered by the police at Eureka. This apparent reluctance by the troopers to charge into the fray has often been cited as proof of their craven and unsavoury natures. Such an image fits perfectly the Eureka legend, in which the police are universally reviled. Such partisan attitudes are not useful when assessing the actual role played by the police.

There are quite reasonable explanations why the police did not charge, sabres drawn, into battle, and so suffered no serious casualties during the storming of the stockade.

It must be remembered that they were not soldiers. In modern parlance, it was not in their job description to place themselves in a situation that required them to accept the distinct possibility of death at the hands of their enemies. Nor would they have been required to do so. The troopers were never intended to storm the stockade. Their task, explained by Thomas in his after action report, was to threaten the stockade's flank and rear.

In military parlance, threaten means to distract an enemy, or restrict his tactical options by one's presence. It does not necessarily mean to charge into the fray unless an opportunity to do so presents itself. Such an opportunity did not occur during the battle for the stockade. Faced as they were by much more determined resistance than anyone on the government side expected, it would have been natural and logical, not craven, to allow the military professionals of the army to bear the brunt of the battle.

There were also practical matters that militated against the police charging headlong into the fray. The troopers were mounted, while the wall of slabs they faced was of sufficient height and sturdiness to make it difficult, if not impossible, for mounted men to cross, especially against armed opposition. To ride up to those slabs with no hope of crossing them, while they were still defended by determined men with firearms, would have been foolish indeed. To hang back and fire from a distance was the sensible thing to do in such circumstances.

The smoke from the firing and the poor pre-dawn light would have reduced visibility. Simply being able to spot, and thus have any chance of hitting, the defenders would have been a major problem for the police. There were 20 to 30 tents inside the stockade.[11] These, along with the slab walls, provided concealment and cover for the insurgents. Numerous accounts from police and soldiers refer to seeing insurgents run between or in and out of these tents. Constable Thomas Milne, for example, saw 'a number of men running from a tent towards where the military were stationed'.[12]

Constable John Badcock also claimed he saw a man he assumed to be Carboni 'running around the corner of a tent'. Badcock aimed his firearm at the fellow, but it misfired.[13] Such were the fleeting views of their enemies granted to the troopers. To rush forward against such an armed and elusive foe would court disaster. Unable to rush among the insurgents, unsure what lurked among the tents, and taking fire from the insurgents, it was little wonder that the troopers hung back and preferred to use their firepower.

As the light grew and the details of the stockade and its defenders became more apparent, the troopers would have studied the situation before them. It would soon have become apparent that there were very few insurgents defending the part of the stockade they were facing.

Lynch states quite clearly that there were not enough men present to cover the entire face of the stockade, and that the left, by which he meant the western, face was not adequately covered.[14]

No doubt emboldened by their discovery, Furnell's troopers moved forward and intensified their firing. Their shots would have fallen on the unprotected left rear of the insurgents facing the main attack, who were squatting and kneeling behind the palisade. It was that fire, ironically delivered by a force that has emerged from the countless retellings of the Eureka story as the most loathed and despised of all the participants, that broke the back of the Eureka insurgents' defence. The tactically decisive moment in the battle had arrived.

Lynch, sheltering behind his reinforced slab barricade next to Diamond's store, described that moment when it came, 'our left being unprotected, the Troopers seized the advantage, wheeled round and took us in the rear. We were then placed between two fires, and further resistance was useless'.[15] His reference to being placed between two fires is significant. The insurgents now had shots coming at them from their front and from behind.

Soldiers detest having the flank or rear of their formations threatened by an enemy. Such a threat indicates to the unfortunates concerned that the enemy has captured the tactical initiative. Even more, it signals that the line of escape may be about to be cut off. When this is combined with a serious threat from the front, and there seems no way to relieve the situation, catastrophe normally results. Such circumstances can lead to panic in even the staunchest military units.

One example of the consequence of a threat to a flank, from myriads that exist, can be found in the fate of the supremely confident veterans of Napoleon's Imperial Guard in the final grand assault launched by the French at the battle of Waterloo. Advancing against the British line, *La Garde*, finding their way forward blocked by the fire of their enemy and their flanks assailed, fell into disorder and were driven back in disarray. If such a thing could happen to veterans, most of whom had seen some 20 years service in the harshest of campaigns, one could imagine what effect a similar experience would have upon what, for the most part, were the very unmilitary defenders of the stockade.[16]

The Eureka insurgents' position was now tactically untenable. In response, and quite understandably, many decided that discretion was the better part of valour. Abandoning their posts, they turned and ran. Carboni captured that moment well when he wrote that 'a whole pack cut for Warrenheip'.[17] This soon became a mad rush for safety by many, including Vern. It would have been about then that the fire from the stockade should have become desultory.[18] And yet, it does not seem that those facing that fire noticed this occurring.

O'Keefe stated that a great deal of firing was still going on up until the time the order to charge was given, which was after many insurgents had fled.[19] Richards made the same point, recalling that 'on both sides the fire was kept up heavily until the stockade was taken'.[20] Webster, referring to the number of casualties suffered by the soldiers when they began to enter the stockade, went so far as to state that there 'were so many men killed or wounded there in the course of a moment or two, and the men not seeing exactly what it was, hesitated for half a moment'.[21] Webster's account, while perhaps exaggerated by a man not used to seeing such spectacles, indicates that the soldiers still faced an opposition formidable enough to take note of. There is no indication from O'Keefe, Richards or Webster that the fire from the defenders ever became desultory, reflecting the perceptions of men who were actually being shot at, rather than writing later from second hand information.

Encouraged by the flight of numerous insurgents, Thomas ordered up his supports, who had been waiting behind the skirmish line. Reinforced, the soldiers pressed up almost to the palisades. Lynott saw a man armed with a shotgun near some tents to his right. Incongruously, the shotgun-toting insurgent was wearing red drawers, perhaps signifying that he had no time to pull on his trousers before the battle had begun, or that he was Dutch, a nationality who wore red trousers to mark themselves out on the diggings. Lynott fired, but the ball flew wide. Then the order came to fix bayonets.[22]

O'Keefe distinctly remembered that he was standing only a few yards from the stockade, alongside Wise, when this order was given.[23] Hearing the command 'Fix Bayonets', the Redcoats would have reached down to their right hip and slid the 22 inch (56 centimetre) triangular iron blades from their scabbards. A practiced, swift turning movement then fitted the bayonet securely to the muzzle of the musket. With that simple act, the soldiers knew that the time to close with their foes and finish the business once and for all had come. Before that happened, one final volley was fired.

A volley is the simultaneous discharge of a significant number of a unit's firearms at the same general target. It is different to the more individualistic firing delivered by the skirmishers. There are numerous references to several volleys being fired by the soldiers during the course of the battle. Tuohy, a stockade defender, recalled the soldiers firing three volleys that morning.[24] The first of these, described earlier, was the initial volley ordered in response to the fire received from the stockade at the beginning of the engagement. If Tuohy's recollection was accurate, then one other volley must have been fired sometime during the exchange of fire outside the stockade, before the final volley.

Each volley fired by the men of the 12th and 40th would have sent a shower of lead balls into the slabs and carts of the stockade, as well as the flesh of any insurgents unlucky enough to be exposed. Many of those balls would have flown wide and, as we have discovered, the impact of those that did strike home might not be as catastrophic as one might imagine. The likelihood of severe injury or death would, however, be high, especially as in most cases the only uncovered part of the insurgents sheltering behind the slab barricade would be their heads. The third and final volley was definitely fired just before the order to charge the stockade was given. Carboni described its effect as mowing down all those who had their heads above the barricades.[25] Despite Harris' earlier observation, quite obviously not all the insurgents were being 'shy' at that moment.

Insurgents Michael and Patrick Canny, Teddy Moore, and Patrick Gittings, were firing at the soldiers from dubious shelter behind an upturned dray. Michael Canny, whom we have previously met describing the wound he was now about to receive, saw Wise fall, probably for the first time. He fell very near to, or on, the palisade, which identifies the fusillade Canny and his comrades then received as the one delivered just prior to the order to charge being given.

When the volley was fired, Gittens and Moore fell. Canny assumed both were killed, but Moore was found mortally wounded by Carboni after the battle, and eventually died from his wounds.[26] At the same time Michael's brother Patrick fell to the ground with his shin smashed by a musket ball. Michael, despite being shot through the arm and breast, turned and fled.

Having fired their volley, their muskets now unloaded, and with bayonets fixed, the soldiers surged forward. About this time Wise, already wounded in one leg (at which he had supposedly quipped that his dancing days were now ruined) was shot again.[27] Interestingly, Lynott recalled that Wise was already lying wounded when he was hit the second time, which conflicts with the image of him leading his men over the palisade, sword in hand.[28]

Regardless of Wise's actual circumstances when he was hit, this time the head of his tibia and fibula in one leg were shattered, inflicting a wound that ultimately proved mortal. Private John Sullivan of the 40th saw Wise on the ground, and moved over to him.[29] Two soldiers, perhaps O'Keefe and Sullivan, lifted him under each shoulder and dragged him to safety behind a nearby mullock heap.[30] Official accounts credit Wise with cheering his men onward as he lay wounded. This may have been so, but no first hand accounts support the story.

Who shot Wise was never determined. Several years after Eureka, Madocks, who claimed to have been inside the stockade during the fighting, was reading a book published in 1856 that included an imaginative account of Wise's wounding. Madocks, incensed by the account, wrote in the margins '[p]ositively false – Patrick Murphy shot Wise'.[31] However, the names of those known to have been inside the stockade, and lists of those arrested at Eureka, do not include a Patrick Murphy. This does not, of course, mean that he did not exist; it is, after all, a common enough Hibernian name.

We have already considered Ferguson's claim that Burnette shot Wise, and explained why that could not have been the case. Another account of Wise's wounding was given by Pierson, who claimed that when Wise was shot, several insurgents were prevented from killing him by the intervention of another American insurgent.

Pierson, despite being extraordinarily passionate in his 6 December diary account of Eureka, did not claim to be at the stockade. He described how the mortally wounded Wise, in appreciation of the role played by the American in saving his life, requested that the man, and another American, be released from custody.[32] It is difficult to reconcile the events Pierson described with the account of O'Keefe, who was within a few feet of Wise when he was hit. O'Keefe made no mention of any such good Samaritan intervening to save Wise. Likewise Sullivan, who stepped over to stand near Wise when he went down, made no mention of anything like what Pierson describes. The accounts of the two soldiers are the more believable.

Wise, leading his men forward, would have been in the company of soldiers when he was wounded. It would have been most unlikely that any insurgent could have approached to render assistance without being shot or bayoneted. What Pierson's account does do, though, is emphasise once again the presence of Americans at the forefront of the battle at Eureka.

Official suspicion for the wounding of Wise fell most heavily on the African-American Joseph, who was only a few paces from Wise at the time. He had certainly fired one barrel of his double-barrelled gun in Wise's direction, but he was with two or three other men, all of whom were also firing.[33] Perhaps Joseph was blamed with shooting Wise because as an African-American, he was more conspicuous in the heat of the moment than his comrades. He certainly seemed to have attracted the ire of the soldiers, and according to one account barely escaped being shot out of hand when he was captured.[34]

The order 'Charge' had been sounded. Wise had fallen, and bullets were still flying thickly. The soldiers hesitated for a moment. Thomas, who had dismounted, no doubt sword in hand, called out something like 'Fortieth, follow me!'[35] Encouraged by his example, the Redcoats, their momentary hesitancy forgotten, cheered and surged forward.[36]

The soldiers clambered up and over the palisade. The slabs were tightly packed, leaning out towards the attackers, and sunk someway into the ground. They made a sturdy obstacle. The men of the 12th and 40th scrambled over as best they could, some jumping up on top of the slabs, some stumbling and falling.[37] Those defenders still holding their ground continued to fire at their attackers. Paul fell severely wounded with a bullet in his hip.

Private James Gore of the 40th, out on the right of the line, was one of the first over the palisade. Having already fired his musket, he was confronted by a man whom he thought to be Carboni, armed with a pike.[38] Gore was genuinely mistaken in identifying Carboni as the man coming at him with a levelled pike in hand. There was another man inside the stockade, a Scot named John Robertson, who so closely resembled Carboni that many later mistook his corpse for the Italian. Carboni himself attested to the resemblance, when he saw Robertson's remains after the fall of the stockade.[39] Gore's mistake was part and parcel of the confusion of battle. With the possibility of imminent impalement confronting him, Gore turned and leapt back out of the stockade. His pike-armed adversary pursued him for a brief moment, but seeing the mounted troopers arrayed out in the open, retreated.[40] Other soldiers crossed the barrier, their numbers rapidly multiplying as they climbed, stumbled and lunged their way inside the stockade.

The conflict now became one of desperate individual combats. Ferguson recalled the fighting when the soldiers crossed the palisade as savage and 'the most exciting time of my life'.[41] During this wild melee, an extraordinary act of bravado and courage occurred. Those men from 'the division having revolvers' rushed right up to the soldiers, no doubt hoping that by doing so, they could make their remaining shots count.[42] These revolver-armed men were most probably Californians, as they formed the only ostensibly revolver-armed division within the stockade. Such behaviour would not have been out of character for them. The sheer reckless audacity of the act was very much in keeping with their typical obstinate spirit of defiance towards obnoxious authority. 'I'll die before I run' were the words of a ballad popular in California during the gold rush era,[43] that expressed a characteristic trait noticed by many who knew the Californians working the Victorian goldfields, be they native-born Americans or Whitewashed Yankees.[44]

Writing of Americans at Ballarat in 1853, Gold Commissioner C. Rudston Read described them as men who were 'awkward customers to deal with' if others had made themselves obnoxious to them. An American merchant at Ballarat explained that his countrymen were 'dreaded more than any other class of men on the diggings, not because they are much disposed to fight, but when they fight they fight with a purpose'. [45] G.H. Dawson, describing those Americans he met at the Bendigo diggings, noted that they 'would lay down their lives for a mate'.[46] Policeman John Sadleir recalled that the Americans he met on the Ballarat diggings were 'perhaps not liking restraint overmuch, certainly not bearing it with the quiet patience of the ordinary Britisher'.[47] It would be second nature for such men to take the fight directly to their foe.

By rushing forward, the revolver-armed men were said by a correspondent for the *Argus* to have caused 'irredeemable confusion' to the defenders of the stockade, as they could not fire their weapons at the soldiers.[48] The revolver-armed insurgents may well have masked the fire of their comrades to their rear. However, the correspondent protests too much. In reality, it is hard to see just how much more confusion could be inflicted on those who remained to defend the stockade, at a time when the defence had all but collapsed.

Their impetuous courage, however, must certainly have spelt doom for at least some of them. Yet no Americans were officially recorded as killed at Eureka, although at least one, perhaps two, were recorded as wounded.[49] Given their prominence in the battle, at least some Americans must have died at Eureka, and there are indications that this was so. The 19 February 1855 edition of the *Daily Alta California*, San Francisco's newspaper of the day, reprinted a front page report received from the Melbourne correspondent of the *Sydney Morning Herald* that claimed that 'a number of Americans joined the insurgents: that they were the most forward in the defence of the stockade and that ten of them fell'.[50]

One insurgent known only as Black Jack, who came to Victoria aboard a ship from New York and thus was probably American, was said by an American miner to have been killed at Eureka. In an interesting footnote, the report of his demise adds that he 'however was a man who could be well spared'.[51] In 1876 the Victoria Police received a letter from California enquiring about one A.A. Bailey, an American whose family were concerned that he may have been killed at Eureka.[52]

The strongest piece of circumstantial evidence that Californians were killed at Eureka was the significant number of unidentified corpses buried after the battle. Three batches of unclaimed and unnamed bodies were buried. One group of five was buried by Inspector Foster, a group of ten by Evan Watkins, an undertaker, and a further group of six by the police.[53] It is highly likely, given their prominence in the battle and their intimate proximity to the onrushing tide of bayonet wielding soldiers that at least some courageous revolver-armed Californians were among the unidentified dead.

About this time, Lalor fell. He had been standing on a pile of slabs that covered a disused hole. Lalor certainly did not lack courage, nor resolve. His position, as befitting a commander-in-chief, was an obvious and exposed one from which he could see much of what was happening around him, and issue orders. Yet every moment he spent there toyed with fate. Lalor gestured for the men at the stockade wall to withdraw into to the cover of the shepherd hole rifle pits, implying that those were still in the possession of the Californians. He had aimed his double-barrelled pistol in the direction of the soldiers, when a shot hit him in the top of the left shoulder.[54] Carboni described this as a chance shot, although it was later claimed that the policeman Lawler had fired the shot.[55] Lalor, hit by one musket ball and two smaller bullets, staggered, dropping his pistol. Recovering himself, he leaned over, his left arm now useless, to pick up his pistol with his good right hand.[56]

Ferguson, who along with at least one other Californian was near Lalor when he was hit, left a vivid account of the moment, stating that the:

> soldiers were in among us. Lalor was shot in the arm, and Hull pulled off his necktie and we wound that around it. He was bleeding profusely and before we were through he fainted from loss of blood. We put him in a shallow hole and covered it with some slabs.[57]

Once again, however, the Eureka story mill produces a rich bounty of alternative tales. One otherwise uncorroborated account has Lalor crying out just before he collapsed, '[g]et away boys, quick as you can, the stockade is taken'.[58]

O'Brien, recalling his role in the defence of the stockade, described how he was with Lalor and Black during the battle, but managed to escape. Before this happened though, he saw Lalor surrounded by 'a dozen soldiers' who were about to bayonet him when the Roman Catholic priest Father Smyth intervened, and crying out '[f]or God's sake leave him to me', saved his life. O'Brien claims it was Father Smyth who got Lalor into a six foot (1.8 metre) deep hole, and left him there, before returning later to spirit him away.[59]

O'Brien's recollections, recorded in his later years, are not to be taken seriously. In 1855 O'Brien swore under oath at the State Trials that he was nowhere near the stockade during the battle. Father Smyth, in his own testimony at the trials, and in a letter written after the event, made no mention of such an event, or that he was inside the stockade during the battle. He was at pains to mention that he only entered the stockade immediately after the battle, and was prevented by the police from remaining there to minister to the wounded and dead of both sides.[60]

The accounts of Sutherland and Ashburner also contradicted O'Brien. Sutherland's recollections, rendered second hand, made no mention of an intervention by Father Smyth.[61] Ashburner was an eyewitness participant, and described how the wounded Lalor asked to see Father Smyth, this quite obviously having been either before Lalor lost consciousness, or at some later stage when he regained consciousness. As Ashburner knew Smyth, he volunteered to go and get him, which he admitted he was very glad to do, 'to get clear of the place'.[62] It is apparent that Smyth was not present at the time.

However he managed to get there, Lalor lay hidden in his hole, covered with slabs and safe from discovery by soldiers or police. The fighting for the stockade now quickly passed over and around him.

Ross, whose company of insurgents defended the northern part of the stockade, fell soon after. The spirited Ross was well known and respected among the insurgents. He had been chosen as a captain of riflemen, and was very active leading parties of armed insurgents prior to the attack. Two days earlier, Ross had boldly carried the Southern Cross flag at the head of the column of miners as they marched from Bakery Hill to Eureka. Now he stood near the base of the flagpole on which hung the same flag.

With many of his men now dead, wounded, or fled, and the soldiers clambering over the palisade, Ross turned to Ferguson and said, 'Charlie it is no use, the men have all left us'. Not a moment later he exclaimed '[m]y God I am shot'. Ross, hit in the groin, died later in the day. Claims were made after the event that Ross was shot some minutes after he had surrendered.[63] This is not likely.

No mention was made of Ross by any of the soldiers or police who stormed the stockade. He was a prominent insurgent, and would have been well known. That he was not mentioned at all, either as a casualty or prisoner, indicates that he had fallen from sight sometime during the battle. In his memoirs Ferguson, who was with him, strongly rejected as nonsense claims that Ross had been shot after he surrendered, and adamantly insisted that Ross fell at his side during the battle.[64]

With the soldiers surging into the stockade, Ferguson knew that there was no time to tend to Ross. Challenged by a soldier, he had no choice but to attempt an escape. It must have seemed at that moment that the insurgents' fight for the stockade was over. Yet it was not. One more heroic act of defiance occurred before the stockade's defenders finally conceded.

Curtain's pikemen had been posted in the upper part of the stockade, facing the Melbourne road. They held their ground when those insurgents who fled had rushed out of the stockade. As the soldiers came up to, and then began to cross, the palisade, Curtain's men moved towards them. To do so when all around them was collapsing in ruin speaks mightily of those men's courage and determination.

Miller, the 15 year old miner whom we briefly met earlier, describing how the stockade was constructed, was one of Curtain's pikemen. He left a vivid account of the struggle between the pikemen and the soldiers.

Miller found himself squaring off against a burly soldier of the 40th. With each thrust and parry of pike and bayonet, a fierce, desperate, struggle ensued. Miller recalled that, because of its length, the eight foot (2.4 metre) long pike was superior to the musket and bayonet for such work, falsifying the notion, so often repeated, that the pikes were flimsy and inadequate weapons. Fighting desperately, he soon became aware that the soldier, who was using his bayonet in a masterly fashion, outclassed him. In an act that showed both his inexperience and the impetuosity of youth, Miller dropped his pike and took hold of the soldier's musket. The soldier, who was much stronger, wrenched his weapon free and clubbed Miller on the head with the butt of the musket, knocking him senseless for a few moments.

With shots from police revolvers and carbines reducing their ranks, the pikemen stubbornly stood their ground, but they could not resist the soldiers' bayonets for long. With their numbers decimated, and all hope gone, those who still could turned and fled.

Recovering consciousness, Miller looked about and saw that everyone around him was attempting to flee the stockade. Deciding to join them, he got to his feet, but a shot struck his left leg and he fell, unable to get up again. A mounted trooper whom Miller called McIver rode up. Leaning over the neck of his horse, the trooper lunged at Miller with his sword. Miller threw his left hand up to catch the blade, but the sword pierced straight through it. When the trooper pulled back, the sword's blade severed Miller's thumb and nearly did the same to his forefinger. Just as he was about to lunge again, an insurgent running past lashed out with a pike. Struck on the neck, the trooper fell from his saddle.[65] The only policeman recorded wounded at Eureka was named McIvor.[66] The similarity between the names McIver and McIvor, and the fact that Miller's attacker was knocked from his horse, strongly suggests that he and the wounded McIvor are one and the same.

Throughout their fight, the dour courage of the pikemen was incontestable, Neill recalling that when the bayonet and pike went to work. 'The diggers fought well and fierce, not a word spoken on either side until all was over'.[67] Thomas O'Neill, a native of Kilkenny, exemplified the spirit of the pikemen that day. Despite having both legs broken and a musket ball in his body, he would not surrender. Whirling his pike about his head, he held off any soldier or policeman who attempted to disarm him. They were unable to take O'Neill alive in the face of such defiant fury, and he was shot dead where he sat.[68]

The losses among the pikemen must have been significant. Several sources specifically mention them as the hardest hit of all the groups within the stockade.[69] This may or may not have been so, considering the numbers of unknown dead among the insurgents. Other companies, for example the revolver-armed men who ran up to the soldiers, may have suffered just as heavily. Yet it does not really matter which of the companies of defenders suffered more than any other. Nothing can detract from the truly heroic stand made by Curtain's pikemen that Sunday morning.[70]

From a purely military viewpoint, their stand was a romantic and futile gesture. Curtain's pikemen, although fighting ferociously, inflicted little real harm on their adversaries. Only one soldier received a mortal wound from a pike, at an undetermined time, and only two other soldiers and McIvor were reported as suffering wounds that were not gunshot related. It is inevitable that there would have been soldiers lightly wounded during their fight with the pikemen. Yet the cuts, abrasions and bruises inflicted would have been considered insignificant, and passed without comment. What the brave pikemen did achieve, for a few moments, was to distract the attention of many of the soldiers and police then swarming into the stockade. This would have allowed other insurgents to flee without encountering Redcoats and troopers.

It could be argued that this was small compensation for the expenditure of so many lives. Yet such an interpretation, while an accurate analysis of harsh reality, is too narrow a view. It is as a heroic myth that the self-sacrifice of the Eureka pikemen can be best understood. Their doomed heroism fits perfectly the role of martyrdom in a righteous, but failed, cause, just the sort of mythology that has received enormous sympathy in countless retellings of the Eureka story.

Even after the inevitable romantic embellishment, the pikemen of Eureka have become the perfect examples of heroic defiance against overwhelming odds. Sacrificing their lives to protect their mates also fits well into the powerful Australian cultural ethos of sturdy masculine interdependence and loyalty, characterised later by both the bush and Gallipoli legends. Such an iconic image creates the prism through which we can view the entire battle at Eureka. Seen in this light, the brave but doomed fight by Curtain's pikemen becomes a manifestation of the passionate resolve not to surrender to the imposition of repugnant authoritarianism, the very stuff that has made Eureka relevant to each succeeding generation of Australians.

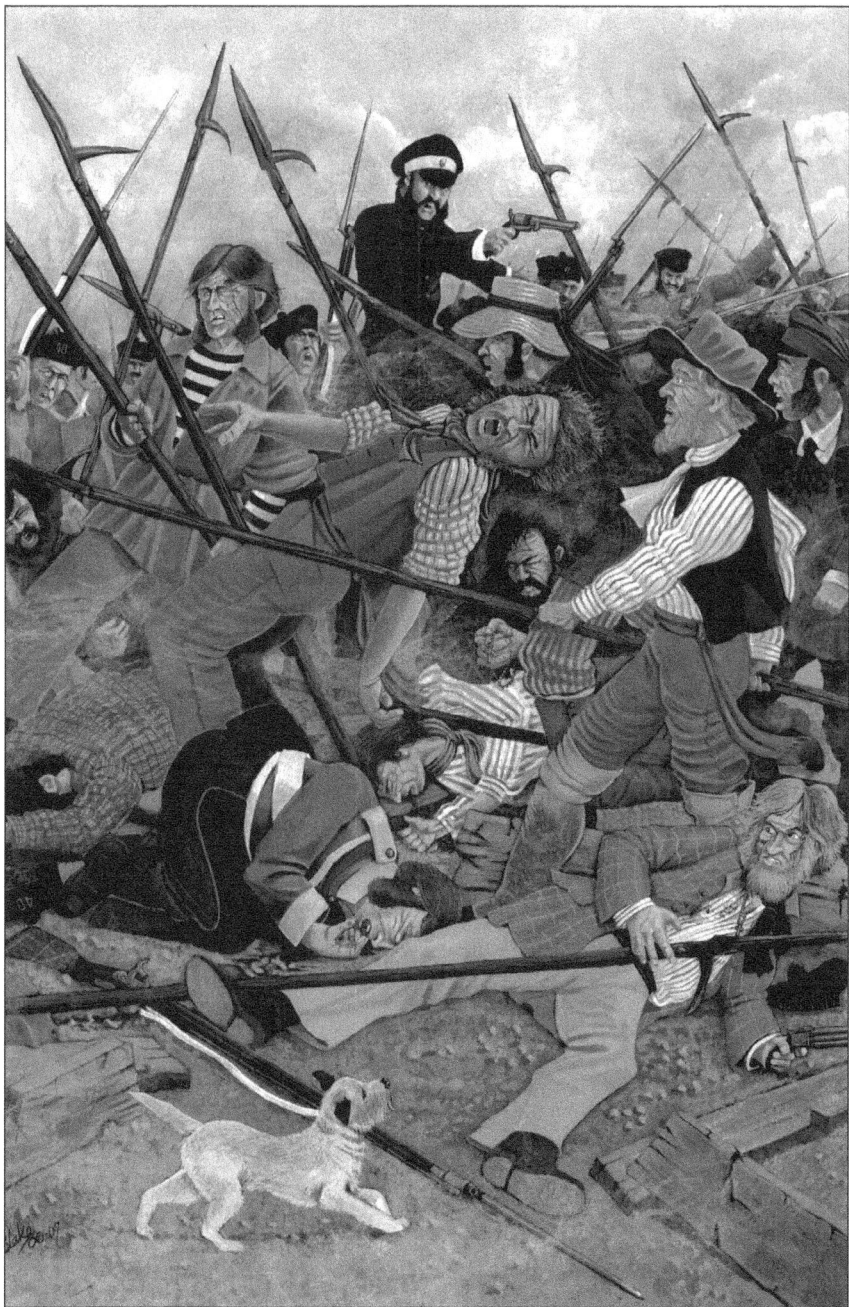

The Pikemen at Eureka. Patrick Curtain's pikemen fought heroically in a futile attempt to stem the tide of Redcoats as the soldiers surged into the stockade. Some suggested that the pikemen suffered the most losses of all the insurgent companies present within the stockade that morning. (Original panting by Gregory Blake)

Some time before or during the final assault, Thonen fell. When Carboni found his body after the battle, he described him as having 'his mouth literally choked with bullets'.[71] The circumstances of his death remain unclear. The Ballarat correspondent for the *Argus* recounted the death of 'one lad, an Italian who was well known on the diggings as a seller of lemonade'. The correspondent's report described the lemonade seller gamely firing the last shots from his revolver before he died.[72]

Although not a lad, nor an Italian, this story, no doubt romanticised, might have described Thonen's death. He was well known on the diggings as a seller of lemonade, the only one specifically named in the context of Eureka. Another account puts Thonen standing on top of the palisade encouraging his men, when he was hit in the face by two bullets and died instantly.[73] There is no way to determine the actual circumstances of his death. Whatever the cause, each story creates a graphic image, very much in keeping with what could be expected to have been happening inside the stockade.

It was not just miners whose lives were in danger. Women and children were also present in the stockade. Twenty-six year old pregnant Glaswegian Mary Faulds cowered inside the tent she shared with her husband Matthew. As bullets flew all about, Matthew rolled logs to each side of Mary, who lay down between them. At some time, almost certainly after the defence of the stockade had collapsed and the palisade breached, a mounted police trooper slashed open the canvas of her tent with his sword. When he saw what lay inside, he rode off. Family legend would later have it that the arrival of Mary's baby was hastened by the noise, smoke and mayhem occurring all about her that morning.

Sixteen-year old Anne Duke had helped sew the Southern Cross flag. She and another woman, known only as Mrs Parker, also had a tent in the stockade. Pressing themselves down flat behind some logs in front of their tent, both women sheltered from the fire. As they lay there, bullets cut through their tent, struck utensils and riddled a box holding clothes. Seven-year old Catherine Donnelly fled, along with some other children, and became lost during the affray. It is not certain if they were inside the stockade or close to it during the battle. She and the other children found safety with some Aborigines. Eighteen-month old Elizabeth Amies' family lived in a tent inside the stockade at the time of the disturbances. She may have been there during the attack.[74] Somewhat naïvely, Mary Jane Humphris went to the stockade with some other women to see what was happening. When the shooting began, she sheltered behind a water barrel. A bullet cut through the brim of her sunbonnet at one time she looked up.[75]

'Old Waterloo' Allen testified that three tents were next to his inside the stockade. The first had in it a man, wife and six children, the second a man, wife and three children, and the third a man with five children. Presumably, these families left prior to the fighting, fled early in the action, or, like others trapped by the fighting, did their best to stay under cover and out of sight. Allen stayed in his bed during the fighting, remaining there until forced out by Carter.[76]

The Redcoats were now across the slab palisade and fanned out among the tents inside the stockade, to root out those defenders who remained. Foot police had rushed ahead and joined the soldiers. John King ran to the base of the flagpole on which had been nailed the Southern Cross flag of the insurgent miners, and began to climb. While he was doing so the pole broke, split and weakened by hits from the multitude of musket balls fired at the stockade.[77] King grabbed the Southern Cross flag as a trophy, tearing it loose from its now shattered pole.[78]

Insurgents who had either no means of escape, or had deliberately decided to stay and fight it out, were still firing from various places. James McFie Campbell, a West Indian, ran between two tents. He had a pistol and rifle with him, and fired at some soldiers about ten yards (nine metres) from the stockade.[79] Lynott saw Manning standing by a tent and firing at the soldiers.[80] He also recalled shooting coming from a tent he described as the blacksmith's tent.[81] Hugh King described fire continuing from 'several posts' within the stockade.[82] Richards recalled several shots fired at the soldiers from tents.[83] Describing the shooting still coming from the some of the defenders, Hegarty stated that there 'were shots fired out of tents, bullets I believe; I could hear them whistling by'.[84] One tent more than any others seemed to attract the attention of the soldiers and police.

The Guard Tent was the largest within the stockade. According to the State Trial plan, it was situated on the northwestern side of the stockade, adjacent to the point of attack. That it held a special significance for the insurgents seems certain. Richards saw armed defenders run towards it when the shooting began.[85] Sadleir said Carter had mentioned in conversation the insurgents running pell-mell into what he called the 'drill tent'.[86] Some 18 to 20 dead and wounded insurgents were found there after the stockade fell.[87] This represented a significant percentage of their losses. It is therefore worth considering the fatal attraction the Guard Tent had for the Eureka insurgents.

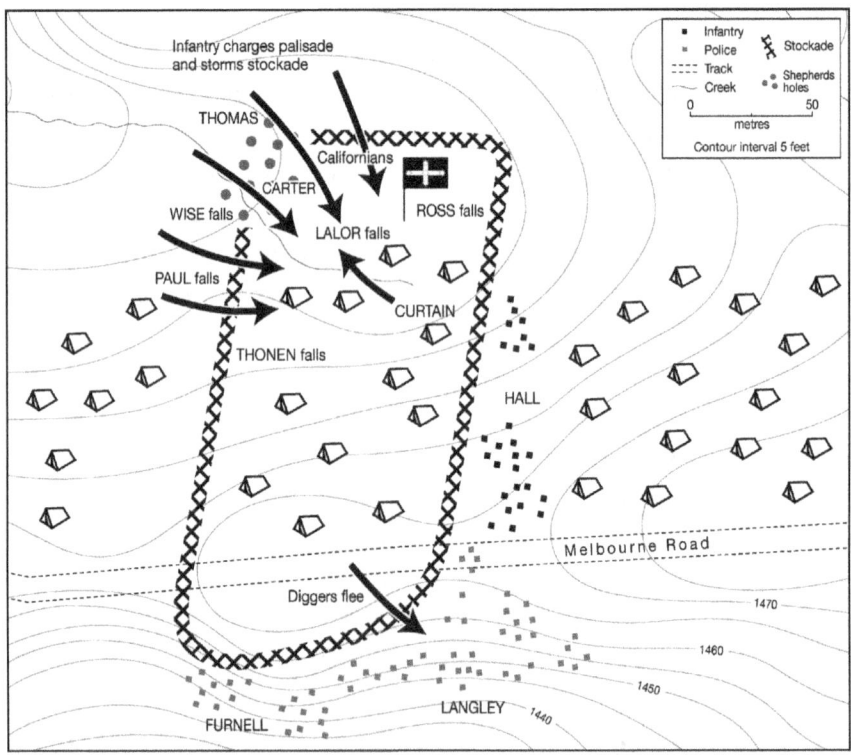

The Battle, phase three: Storming, Rout and pursuit. When the soldiers storming the stockade from the front and police behind them most of the stockade's defenders fled. When this occurred some of the revolver armed Californians charged at the soldiers. Patrick Curtain's Pikemen also attacked and engaged the infantry in a fierce contest, suffering heavy losses. While the infantry dealt harshly with any insurgents they could catch the mounted police and mounted 40th pursued the fleeing insurgents.

Significant stores of weapons and ammunition were found there, attested to by both Richards and Carter. Had the insurgents seen running to the Guard Tent intended to use these? Or did they carry those weapons into the tent with them? The correlation of the number of dead and wounded and the number of weapons found suggests that the latter might have been the case, however, Carter's description of the weapons found there as 'stands' indicates that they were stored there.

The insurgents who had ensconced themselves in the tent found themselves the target of a heavy fire from the soldiers. Sadleir recalled that Carter described what then happened as a tragedy. The insurgents, clustered together, were shot down as the soldiers fired into the tent. Despite having slab sides, the tent would not have offered more than token protection to anyone sheltering there.[88] Luckily for both Joseph and Manning, who were captured when they came out of the Guard Tent, Carter, seeing what was happening, signalled to the officer in charge to order his men to cease firing. Carter was of the firm opinion that if he had not done so, scarcely one of the miners gathered there would have survived.[89]

Those insurgents who still could now fled for their lives, running 'all over the place' in the words of miner Richard Smith, who had been asleep when the fighting began.[90] James Beattie, from Cumberland, England, was one who ran. Climbing over the stockade, he found himself confronted by mounted men of the 40th and police troopers. Dropping a large horse pistol he was carrying, Beattie went down, either kneeling or falling.[91] Handcuffed he was led around to the front of the stockade by Sergeant Patrick Riley of the mounted 40th. This was about the time firing inside the stockade ceased. Fearing for his life at the hands of his captors, Beattie pleaded with two other men, who may also have been prisoners, not to leave him.[92] Beattie survived his ordeal and was one of the 13 men charged with treason and committed for State Trial following Eureka.

Tuohy had a double-barrelled shotgun in his hand as he fled. John King called for one of the mounted 40th to apprehend him.[93] Corporal William Richardson gave chase and Tuohy, with no hope of avoiding capture, threw down his weapon and fell to the ground.[94] Henry Reed, or Read, was seen by Furnell inside the stockade as the firing all but ceased. Furnell called on him to surrender, but went unheeded. Even when Furnell fired two shots he refused to come out, and a trooper had to go into the stockade and apprehend him.[95]

In another part of the stockade, 34 year old John Rodan had been slightly wounded in the right shoulder. He crawled under a dray in an attempt to find a way out of the stockade, but was seen and apprehended by Kossak.[96] John Francis Romeo, a native of Corsica, stood his ground, but clutched his arms nervously as he did. Looking right and left, he searched frantically, but in vain, for a place to escape.[97] John Phelan, a friend of Lalor, was armed with a sword. He found himself in a desperate situation as the soldiers swept over the defenders. A shot whipped close by, missing him. He turned and tried to flee. Trooper John Culkin, who had fired the shot, lashed out, striking Phelan with the flat of his sword, the blow preventing him from escaping. Lieutenant Hall saw the struggle between the two men. No mention is made if Hall was mounted or on foot when he jumped the stockade, but he seized Phelan, disarming and handcuffing him.[98]

In a desperate act, Michael Dynan ran through the shambles with a child in his arms. He later claimed that he had been in the stockade to visit his sister, and was released without charge. The identity of the child was never revealed.[99]

Not all the Eureka insurgents were easily run to ground. John Keenan had been shot in the leg early in the battle. Other insurgents carried him to where some horses were tied up, and concealed him among the litter between the horses and the manger from which they were feeding. Keenan remained undiscovered in his improvised hiding place.[100] Thomas Dignum, one of the four known Australian-born defenders of the stockade, turned on the mounted soldiers who had pursued him and five other fleeing insurgents. Using the pike he was carrying, Dignum struck out at his pursuers. Just what he achieved by this is not clear. William Revell of the 40th, who testified that he was the target of Dignum's lunge, claimed at one time that Dignum's pike stuck a soldier with him, and at another time that the pike was run into his horse's flank. Regardless of what actually happened, Revell pulled his horse up and cut Dignum on the head with his sword, capturing him.[101]

Private Thomas Bodely of the 40th saw William Molloy inside the stockade, and drove him out into the hands of police Sergeant Edward Viret. Refusing to be cowed, Molloy turned on his attackers, telling them that 'it was only the beginning of the row. It would not end there'.[102]

Eureka Slaughter. Charles Doudiet's painting of the battle for the stockade shows the soldiers in ranks in front of the stockade with mounted soldiers to the flank. In his depiction of the stockade he portrays quite clearly the large Guard Tent, where many of the stockade's defenders fell. (Collection: Ballarat Fine Art Gallery)

An insurgent named Bryant, whose given name is not known, jumped over the stockade and struck out at Henry Perry of the mounted 40th with a pike, with no apparent success.[103] Having dropped his pike, Bryant continued his bid for freedom, and ran towards a chimney, perhaps hoping to hide behind or in it, but was caught and arrested by mounted police Sergeant John Gilman.[104]

William Madden was standing next to his friends Patsy Gettings and John Hines when both men fell.[105] Taking his firearm with him, he fled, and as luck would have it soon found himself in the bush outside the stockade. However, his escape had not gone unnoticed. A trooper, sword drawn, pursued, and demanded that he surrender. Turning,

Madden brought his weapon to his shoulder and shouted back '[n]o I won't surrender: get you away to the right or the left before I fire'. The trooper wheeled his horse and galloped off.[106]

One miner, whose name was not recorded, recalled a soldier running a bayonet through his comrade. Shooting the soldier, he attempted to escape, but fell over some slabs and lay still. Waiting until the immediate danger had passed, he got up and tried to escape through the now broken palisade, but instead met five or six other insurgents. Taking shelter behind a chimney, they fired at some troopers leading away a group of prisoners. The troopers immediately turned on their assailants, and in an act that showed commendable restraint given the circumstances, captured them all.[107]

Ferguson had been with Ross when the latter fell mortally wounded. No sooner had that occurred, than a soldier demanded Ferguson surrender, to which, according to Ferguson, he replied 'that I would see him dam'd first'. Jumping over the palisade, he fell, and the same soldier fired, putting a ball through Ferguson's hat. Finding his way blocked by mounted troopers, Ferguson jumped back into the stockade and beat his way through a number of soldiers who tried to stop him. Seeing that the police were still shooting at fugitives, unlike the soldiers, he realised that any further attempt to escape was impossible. With nowhere to run or hide, he surrendered to Carter, noting with some pride that the soldiers had not been able to take him.[108]

One of 'Lalor's Captains', his name never recorded, told how he was sheltering from the soldiers behind Amos' requestioned horse when two soldiers came at him with fixed bayonets. He fired his revolver and hit one, who fell to the ground, causing his comrade to fall back. He then fired again, but, while his men who were inside a tent were cheering him, decided that discretion was the better part of valour, and fled from the stockade. Finding his way blocked by mounted troopers, he ran into a nearby butcher shop and hid in its chimney, were he remained undiscovered.[109]

Joseph was described by contemporaries as 'a tall powerful man of color'.[110] He was seen by numerous witnesses fighting well in defence of the stockade. Captured by the police at the Guard Tent, he was handed over to the 40th.[111] His defiant spirit had not been crushed, however. Some time later, he managed to break free and get over the stockade. He resisted enough for the sergeant in command to order his men to 'shove' him back in with the other prisoners.[112] The soldiers appeared very angry with Joseph, for they suspected that he fired the shot that severely wounded Wise. It was later claimed that only the intervention of officers prevented him being shot by the soldiers when they and their prisoner returned to the government camp.[113]

American James Brown leaped down a mineshaft. A sailor and expert on the rope, he slid down 100 feet (30 metres) to the bottom of the shaft. He was not discovered, and some time later climbed back up, taking two hours to do so.[114]

The soldiers and foot police were now among the tents, pits and detritus of battle littering the interior of the stockade. With their enemies' resistance broken, they set to using their bayonets to put an end to the matter once and for all. Having met stiffer resistance than expected, and with numerous comrades sprawled on the ground behind them, the Redcoats' blood was up. Their killing instincts went unchecked for the moment. Those defenders who had not managed to run quickly enough, or who stood their ground to fight it out, fell under the thrusting, stabbing tide. Carboni described the soldiers 'wonted ardour' as they began to fight their way through the stockade.[115]

Lynch referred to 'the spirit of revenge being uppermost' in his depiction of what he called 'a fierce saturnalia of carnage'.[116] Even though a supporter of the government, Huyghue highlighted the bloodlust of the soldiers when he referred to the defenders of the stockade as being 'literally butchered'.[117] Lazarus, a school teacher from Liverpool, saw the bodies of the dead insurgents left in the aftermath of the Redcoats fury, and described them as '[s]ome shot in the face, others literally riddled with wounds'.[118] He described finding the corpse of a German he had often joked with, 'pierced with 16 or 17 wounds'.[119] One dead pikeman was reported by a correspondent of the *Geelong Advertiser*, who saw his corpse, as having 'three contusions in the head, three strokes across the brow, a bayonet wound in the throat under the ear, and other wounds in the body - I counted fifteen wounds in that single carcass'.[120]

Anne Diamond, who with her husband Martin ran a store described as being 'half in and half out of the stockade',[121] in which the insurgent council of war had met, described how Martin was shot down and struck three times with a sword, as well as bayoneted. It was only a statement of fact when she observed that the soldiers 'were just tearing over the people as if they had no feeling at all', adding that 'they treated the dead bodies very badly'. Adeliza Faulds, whose mother Mary was inside the stockade when it was stormed, recalled her commenting on the 50th anniversary of the battle that the 'soldiers were a lot of flash youngsters who would kill a man as readily as they would kill a chicken'.[122] And kill the soldiers certainly did.

Ned Flynn ran for an old chimney, perhaps hoping to hide inside it and wait out the storm raging all about him, but he never got there. A soldier thrust his bayonet into Flynn's throat killing him.[123] A blacksmith, probably Hafele, fell defending himself against Richards. Neill recalled that the blacksmith 'fought well and died gloriously'.[124] Unable to escape as quickly as he might, Patrick Callinan was wounded by a bayonet between the shoulder blades and under the left breast.[125]

One unnamed insurgent lay desperately wounded, but still had enough left in him to raise his rifle and aim at the soldiers. A blow to his head from a sword knocked him senseless. He was brought into the government camp after the battle, where he died, with six bayonet wounds in his body as well as other injuries.[126] A miner named Gordon Jukes claimed several men were killed while attempting to run away, and observed that the bodies had been 'badly mutilated with sword and bayonet'.[127] A.W. Arnold witnessed 'one man lying on his stomach, wounded, kicking and throwing his arms about. A soldier was standing over him, and I saw the latter put his bayonet right through the Stockader's back, who kicked no more'.[128]

In such a manner did the reality of war descend upon the Gentlemen Soldiers of Eureka.

The stockade had fallen. Those defenders who had not fled had been captured, shot, bayoneted, or put to the sword. All about lay corpses and wounded men, many horribly injured and pierced numerous times. The soldiers, their frenzy abating and their officers reasserting control, set to rounding up prisoners and handing them over to the police. Women had come into the stockade looking for their menfolk, and threw themselves over some of the wounded to prevent further harm to them. One woman remonstrated with soldiers on behalf of her husband, but was treated roughly. An unidentified officer rode up and ordered that she be allowed to see her husband.[129]

Miners, women, and probably some children, who had hidden inside their tents, were roused out, although at least one, Mary Faulds, was let be. The soldiers and police then set fire to some tents and many slabs, charring some bodies that lay close by. Some claimed that these men might have been wounded and purposely allowed to lie close to the flames, or even deliberately burned alive by the soldiers and police. Such stories can only be viewed with scepticism. Given the actions they took to restrain their men in the aftermath of the attack, it is hard to believe that officers like Thomas and Pasley would have tolerated the deliberate murder of insurgents by burning them alive.

The army's task was completed. The bugle sounded 'Assembly', discipline reimposed itself. The Redcoats formed ranks outside the stockade, and marched away dragging the Southern Cross flag behind them, leaving the captured ground to the police. On reaching their lines, the soldiers waited for the order to dismiss, and when it came capered like schoolboys as they ran back to their tents.[130]

In 1890 Henry Lawson penned a verse to the insurgents at Eureka that in part went:

But the light of the morning was deadened an' smoke drifted o'er the town

An' the clay o' Eureka was reddened ere the flag o' the diggers came down.

The clay of Eureka had indeed been reddened, and the insurgents' flag roughly torn down. In about 20 minutes of furious conflict the Queen's law, such as it was, had been bloodily reasserted on the goldfields of Ballarat.

Chapter 9
'JOE IS DEAD NOW'

'Massacre at Eureka … cowardly massacre' trumpeted the Geelong correspondent for the *Argus*.[1] 'Deplorable massacres' howled speakers at a mass meeting held at the Collins Street Mechanics Institute in Melbourne on 5 December.[2] Humffray echoed the outrage with a promise to present a true account of the massacre at Eureka.[3] The *Ballarat Times*, in a black-bordered article, announced to the world a spectacle that was 'sufficient to appal the stoutest heart'.[4] The *Gold Fields Advocate* presented its readers with a scene of unambiguous desolation, lamenting that the:

> Eureka goldfield presents a piteous scene. Women roam the camp crying aloud for their men, their children at their skirts wailing for their fathers, and in the midst of the desolate and heartbreaking sight a lone dog epitomises tragedy with a continuous mournful howl.[5]

Pierson described the soldiers and police who had perpetrated the killings at Eureka as 'most heathenish, Bloodthirsty, disgraceful and cruel'.[6]

In following years, stalwart Eureka veterans, such as Lynch, Shanahan, Tuohy and Allan, insisted that most of the killing occurred outside the stockade after all resistance had collapsed. Such were the expressions of public and individual outrage and moral condemnation that accompanied the news of the battle at Eureka.

Contemporary responses like these, over the generations, have provided ample fuel for those who wished to stoke the furnaces of public and private moral indignation. Many accounts written since have reflected such righteous anger. Statements that the 'whole military action was one of deceit and totally lacking in the accepted codes of conduct',[7] and 'the troops brought disgrace on the Military'[8] are typical examples.

While appealing to populist prejudice, such interpretations are misguided. At best, they accept at face value so-called eyewitness accounts of a massacre.[9] At worst, they indicate an appalling ignorance of the military realities of the era. We cannot hope to understand what occurred at Eureka unless we distance ourselves from emotional hyperbole, and consider what does occur in close and furious armed conflict.

The battle for the Eureka Stockade was a bitterly fought and bloody affair. To understand why events unfolded as they did, we must make the effort to climb into the skin of those engaged in the conflict. This means, more than anything, that we need to put ourselves into the position of the soldiers as they rushed over the palisade, bayonets stabbing and slashing against still determined opposition. When we do, we will discover that what happened at Eureka was not a unique experience in a form of combat that has always been notable for its merciless fury and brutality.

There can be no disputing that there was great carnage inside the stockade during those terrible moments when the soldiers fought their way in and set to work with the bayonet. This was to be expected. By its very nature, the use of the bayonet in battle presupposes very close and personal combat, in which spiking or ripping an enemy was the desired outcome of each individual contest. It is not in the character of such combat to confer quarter. It was certainly not in the nature of the mid nineteenth century British common soldier, who had been imbued with the cult of the bayonet, to restrain himself once the man-to-man contest had begun. In those moments, any civil notions of what was moral or immoral became meaningless. This was war at its most brutal and sanguinary.

Molony, in his book *Eureka*, correctly points out that men 'in the fury of battle, commit atrocities which the so called logic of war renders inevitable'.[10] Huyghue, knowing the soldiers of his own era, wrote 'that men when generally let loose upon an enemy are not angels'.[11]

Assistant Colonial Secretary John Moore showed an understanding of the realities of close combat when he observed on 30 December 1854 that restraining the passions of soldiers aroused by bloodshed and the excitement of combat was very difficult.[12]

The official Goldfields Commission report is often quoted as proof that the mounted police, in particular, behaved in a barbarous manner at Eureka. What is seldom quoted are the qualifications offered to explain their actions. In doing so, the Commission stated clearly that 'these were the excesses rising out of duty', and that the Commission would not 'enter under circumstances into questions of individual conduct, reprehensible, it may be, on both sides, during moments of ungovernable excitement'.[13]

Such behaviour is a constant of warfare, and the battle for the Eureka Stockade while it lasted, was conducted as an act of war. To claim, as do those who propagate the myth of a deliberate massacre at Eureka, that the soldiers and police were somehow acting in an extraordinarily bloodthirsty manner, is naïve, and ignores the evidence of centuries of military conflict. There are many examples of what could be expected in such circumstances. At the battle of Fuentes D'Onoro in 1811, the soldiers of the 79th Highland Regiment, having lost their colonel in a desperate charge against the French, showed no mercy to their enemy when the melee began. One senior officer sent to investigate after the incident concluded that such 'was the fury of the 79th that they literally destroyed every man they could catch'.[14]

A private soldier during the First Sikh War observed that shedding blood in battle was simply self-preservation, which he reasoned was the first law of nature.[15] Commenting in his memoirs of his experiences during the American Civil War, an officer of volunteers observed that men 'can, in the enthusiasm of battle, see and take part in the murderous work without realising how horrible it is'.[16] German soldier, essayist, and novelist Ernst Junger, a highly decorated First World War veteran, when explaining what was to be expected from soldiers who were storming a defended enemy position, wrote that:

> the defending force, after driving their bullets into the attacking one
> at five paces distance, must take the consequences. A man cannot

change his feelings again during the last rush with a veil of blood before his eyes. He does not want to take prisoners but to kill.[17]

Australian soldiers during the First World War expressed similar sentiments. One, writing of his experience of hand-to-hand combat at Gallipoli, wrote that 'a soldier does not want any sentiment … the lust for killing seems very strong'. Another reflected that 'the lust to kill is on us, we see red'.[18] One Australian, describing his time fighting the Germans in France, wrote '[s]trike me pink the square heads are dead mongrels. They will keep firing until you are two yards off them & then drop their rifle and ask for mercy. They get it too right where the chicken gets the axe'.[19] The parallel here to the circumstances facing the Redcoats at Eureka, in which insurgents were still firing at six paces distance from their attackers, is obvious.

The bayonet itself is an instinctively horrific weapon. This is despite the reality that it does not inflict wounds any more severe or dreadful than do bullet or buckshot. Despite this, the bayonet seems to arouse an instinctive abhorrence in those who witness its use, or are threatened by it. Even some battle-hardened soldiers have acknowledged an aversion for the use of the bayonet, such as the American who, during the Mexican-American War, described the use of bayonets in battle as 'like murder'.[20] The moral fury at the thought of bayonets threatening the civilian population on the Ballarat goldfield, repeated many times in the months prior to Eureka by the miners and press, was no doubt a reflection of similar perceptions.

It was a brave person who could stand in the face of a determined bayonet assault; and few did so in open battle, when left with any chance of escape. The statistics of battles during the black-powder era show consistently low numbers of casualties inflicted by bayonets.[21] It was, however, in the defence or storming of fixed points, such as the Eureka Stockade, that the bayonet came into its own. The numerous reports of corpses with multiple wounds imply that the soldiers exercised very little discrimination in those heated moments.

Barbaric as such behaviour appears to the untutored eye, it must be viewed in the context of the military culture of the era. The bayoneting of dead, wounded and overwhelmed enemies was not uncommon. It was not, as suggested by some Eureka storytellers, something especially concocted to punish the insurgents at Eureka. Following the battle of Culloden in 1746, the bayonet was used to kill any wounded Jacobite Highlander found lying on the field. Despite occurring over 100 years before Eureka, the harsh dictums that resulted in the bayoneting of wounded Jacobite rebels had changed very little.

In 1804, Major George Johnson drew his pistol and threatened men of the New South Wales Corps to prevent them killing prisoners after crushing the rebellion by convicts at Castle Hill.[22] In 1837 British troops used the bayonet to rout French-Canadian civilian rebels at Saint-Charles sur Richelieu in a bloody encounter, in which between 48 and 150 rebels died.[23] The Redcoats at Eureka would have thought nothing of behaving in such a manner.

Another reason the bodies of many of the Eureka insurgents had multiple wounds is the idiosyncrasies of warfare in the era. In the colonial campaigns fought by Britain's armies against numerous and exotic foes, it was not unknown for enemies to feign death and rise to attack troops in the rear once they had passed.[24] The experience of the 40th, which had seen hard service in Afghanistan and India, or the 12th, which had served on the frontier of Cape Colony, would have developed within both regiments a culture of merciless chastisement of the enemy. In colonial warfare, the dead were assailed, to ensure that they were either actually dead, or sufficiently incapacitated to pose no threat to the attackers. Veterans would have passed this culture on to the younger soldiers. The repeated bayoneting or shooting of foes as they were passed over by successive soldiers in the heat of battle would have been second nature to many of the soldiers at Eureka.[25]

In discussing the behaviour of soldiers during the battle of Waterloo, Keegan makes the point that 'finding themselves intermingled with the enemy, individual soldiers struck out, in drill book fashion at all near enough to threaten them'.[26] It would have been exactly so at Eureka. In those deadly moments, the defenders of the stockade would have been cut down without remorse or mercy. Such behaviour was not unique, and certainly not something especially reserved for the Eureka insurgents.[27]

Fearsome deeds were done with the bayonet in the minutes following the soldiers' entry into the stockade. In a letter written only a few days after Eureka to Emily Gill, in England, Madocks wrote that once 'in possession the soldiers bayoneted all who were not dead'.[28] His reference to 'all who were not dead' is unclear. If he was referring to every miner inside the stockade, he was obviously exaggerating. If he was referring to wounded insurgents, such a fate was entirely likely for some unfortunates in the first frenzied moments following the soldiers' entry into the stockade. The *Argus* also reported that many insurgents had been bayoneted after death.[29]

The degree of carnage that ensued during the battle especially excites accusations of wanton massacre. The loss of so many lives in a relatively confined space, and in such violent circumstances, was a traumatic experience for most who witnessed it. Grim accounts by correspondents and first hand observers such as Lazarus did much to feed the public outrage.

Such condemnations flew in the face of the grim reality of warfare. One does not have to look far in the 1850s to find examples of uncompromising bloodshed in close combat. William Howard Russell, the correspondent for the *Times* of London during the Crimean War, writing at a time contemporary with the events at Eureka, described the aftermath of battle graphically. In November 1854, he described the dead British and French troops who fell at the Battle of Inkerman, writing that they:

> wore terrible frowns on their faces, with which agonies of death had clad them. Some in their last throes had torn up the earth in their hands, and held the grass between their fingers up towards heaven. All the men who exhibited such signs had been bayoneted.[30]

Describing the battle of Solferino, fought in northern Italy between the Austrians, French and Italians only five years after Eureka, the Swiss observer Henri Dunant wrote that the:

> Austrians and allies trampled one another under foot, slaughtered each other on a carpet of bloody corpses, smashed each other with rifle butts, crushed each other's skulls, disembowelled each other with sabre or bayonet. It was butchery.[31]

Although describing combat of a vastly greater magnitude to that at Eureka, Dunant's bleak account captures the gruesome reality of close combat during the era.

The Eureka Stockade was stormed at the point of the bayonet. It would be foolish and naïve to expect such an assault, to capture a position from which the defenders continued to inflict numerous casualties on the attacking force right until the last moment, to be a benign or even remotely civilized affair. Once inside the stockade, the victorious soldiers would have given vent to their rage and, as we have seen, made sure that none of their enemies were left in a fit state to cause further harm to them.

There was also something personal that contributed to the battle lust of the Redcoats that Sunday morning. The wounding of Wise and Paul had inflamed the soldiers' rage. Pasley described how the soldiers also 'hated' the insurgents that morning, because of the wounding of a drummer boy on 28 November, the mortal wounding of Wise during the battle, and a perceived slur to their honour by the insurgents, who had publicly announced that the soldiers would not fire upon them.[32]

Wise was a well-respected officer, and it was not unusual for the men of the 40th to have reacted badly to his wounding. There was an extraordinarily personal, paternalistic, relationship between men and officers within British regiments of the time. Russell recorded that when regiments became mixed in confusion during the Crimean War, the soldiers would not assist or listen to the officers of any other regiment.[33] When Wise was shot, as he was about to lead his men into the stockade, his loss would have been felt as a personal blow by his men.

In a similar fashion, the severe wounding of Paul would have angered the men of the 12th. For the first few minutes after the troops stormed into the stockade, no one opposing them, or even in the vicinity of those opposing them, would have been safe. The Redcoats at that instant would have been in no mood to give their foes the *false touch*.

It took a few minutes for their passions to cool, and even then it needed serious encouragement. Pasley drew his revolver to restrain some Redcoats from killing prisoners.[34] His action is frequently used to imply that the soldiers were behaving with unusually murderous intent. Such an interpretation is mistaken. The soldier's bloodlust was not extraordinary or unique, nor was Pasley's response to it.[35]

There would be times when the necessities of war meant that the soldiers were unleashed for a time. When, however, their task was completed, or they had gone too far, their gentlemen officers restrained them. This is what Pasley was doing when he drew his revolver. Reports of officers preventing soldiers from shooting Joseph and Manning following their capture also emphasise this role.[36] The Redcoats' passions calmed as discipline was reasserted. Whatever hatred they had felt seems to have dissipated relatively quickly, with one insurgent commenting later that not a word of reproach against the defeated insurgents came from the soldiers, with the exception of one officer who called them a 'Vandemonian looking crew'.[37]

Another aspect of the Redcoats' behaviour was blatant theft from the defeated insurgents. This is rarely mentioned in modern accounts of the battle.[38] From the point of view of the outraged miners and the law, there would seem to be little to mitigate the actions of the soldiers. However, such criticisms again do not reflect the military realities of the 1850s.

In the reconciliation of expenses incurred for troops and police at Ballarat, the pay of the soldiers is recorded as 1/6 per day.[39] In one month, a soldier would make slightly less than £2/10/-. By comparison to others on the diggings, their largesse was meagre indeed. John Joseph had on him £7/-/- when captured, three months' pay for the average Redcoat at Ballarat.[40] A boy aged 16 who was taken prisoner possessed a money belt with £6/-/- in it.[41] Carboni claimed to have had money on him when he was arrested.[42] Ferguson admitted that he had $50.00 on him (presumably American dollars), but hid it in the lining of his hat, and the soldiers did not find it.[43]

The British army had a long history of looting in times of war. The army encouraged such behaviour, and appointed prize agents to collect and distribute prize money gleaned from the auction of looted goods. It was an offence not to surrender to the prize agent any loot a soldier found. During the Napoleonic Wars, numerous examples of looting by the British army occurred, the most notorious being after the fall of the

French held Spanish towns of Ciudad Rodrigo and Badajoz in 1812, where British soldiers ran amok.[44]

During the campaigns in northern India, in an era much closer to Eureka, a great deal of looting occurred. Following the battle of Ferozeshah in 1845, victorious British troops seized the opportunity to load horses and bullocks with loot seized from the camp of the defeated Sikhs.[45] After the battle of Gujerat in 1849, soldiers looted so enthusiastically that, according to one participant, they 'made what we could and did very well'. Following the capture of Lucknow in March 1858, treasures to an estimated value of £1.5 million, not at all the true worth of the actual plunder, were looted and auctioned after the event.[46]

For the chronically destitute common soldiers at Eureka, the situation they found themselves in following the fall of the stockade would have presented a golden opportunity for some personal gain from the hazards they had faced at the hands of the insurgents. Only a man of the most saintly disposition could have resisted the temptation, and such men would have been thin on the ground within the ranks of the 12th and 40th that Sunday morning.

Ferguson mentions seeing soldiers putting all the money they found on the prisoners into their pockets as they 'searched' them.[47] Joseph, the unnamed teenage boy, and Carboni all had their money taken from them. Pierson relates how three soldiers set upon a wounded insurgent whom he knew. While two knelt on the unfortunate man's chest, the other went through his pockets and robbed him of his money.[48] Pierson did not see this happen himself, but it is entirely plausible that such an event occurred. Kennedy O'Brien, who was not even in the stockade, was robbed of a note valued at £20/-/- and a small nugget of gold by soldiers who arrested him near the Catholic Chapel.[49]

Many canny methods were devised by the soldiers to hide their ill-gotten gains, such as concealing money in the waists of trousers, or in boots.[50] It can be assumed that this is where much of the Eureka insurgents' looted funds disappeared. Driven by the exigencies and traditions of war, it was not extraordinary that soldiers of the 40th and 12th would take the opportunity to enrich themselves at the expense of their defeated foes.

That the battle at Eureka was bloody and resulted in many horrific deaths is beyond dispute, however, it was not a massacre. What took place was a short, vicious military engagement, a battle and act of war in every respect. This was the reality, and the context in which judgments of what happened must be made.

One of the most frequent arguments that a massacre was perpetrated is based on the supposed unconstitutional nature of the attack. As with many interpretations of Eureka, this is incorrect. The law as it stood in 1854 gave no discretion to authorities on how they should act in the case of an armed challenge to the authority of the crown. Such defiance was to be met forcibly, and if necessary with deadly force. Even if the insurgents had not fired the first shots, killing and wounding soldiers, and thus transformed the confrontation into a military engagement, the situation still would have required firm action by the colonial authorities.

Even the most disinterested observer at the time could not fail to understand that, for all appearances, the Eureka miners had organised and armed themselves for war. Allan admitted as much when he wrote that it was immaterial who was to blame for the battle that erupted, as both sides were under arms and intending to fire on each other.[51] The insurgents had placed themselves beyond the pale of British constitutional protection, and when the military engagement began, a state of war existed in the eyes of the law. In such circumstances, all civil restraints were suspended.[52] This state would have remained until the enemy, in this case the Eureka insurgents, were overcome. Once such a conflict began, it took on a dynamic of its own, and was fought to a conclusion.

Claims by participants such as Lynch that, once resistance within the stockade collapsed, a truce should have been proffered, are unrealistic nonsense. On the other hand, Tuohy astutely observed that 'anything was fair in war time, and we should take what we got and bear it'.[53] The ultimate military realist, American Civil War General William Tecumseh Sherman, described war as 'cruelty, and you cannot refine it'.[54] In Shakespeare's play Henry V, the king beseeches his men before the breach at Harfleur to cast aside the modest stillness and humility of peace and to '[d]isguise fair nature with hard-favour'd rage'.[55] The actions of the soldiers and police engaged in storming the stockade reflected just such a universal experience.

How many died at Eureka? Despite claims by various participants, and subsequent uncritical repetition of those claims by Eureka storytellers, the actual losses suffered were far greater than ever admitted by either side.

The 16 officially recorded military casualties from Eureka are supposedly well known from the list prepared immediately following the engagement, but this is neither a comprehensive nor an entirely accurate record.[56] Privates John Hall and Denis Brien died at the time of Eureka. Neither appeared on the official casualty lists prepared at the time. Brien died on 3 December, the day of the battle, while Hall died in hospital on 31 December 1854.[57] Another soldier, Private James Hammond, was not listed as being at Eureka, but was recorded as having died on the return journey to Geelong after the battle, however, the cause of his death was not recorded.[58]

One explanation for this discrepancy is that the official casualty list was compiled from the report the surgeon made immediately following the battle, so that it could accompany Thomas' account of the action despatched to the governor that day. In the haste to produce the required document, it appears that Brien was omitted from the list, and even though Hall's and Hammond's deaths may or may not have been a consequence of Eureka, all were subsequently ignored as Eureka casualties.[59] There is anecdotal evidence to suggest that the records kept by the 40th Regiment at Ballarat were not as rigorously maintained as they could have been. In the muster rolls for the period, the 40th listed no men being hospitalised. However, similar records kept by the 12th listed numerous men in hospital.[60] This implies some rather sloppy work by the regimental clerks of the 40th. One cannot but wonder what else they neglected to record.

One feature of the military casualty list that is apparent is that it reflects only those soldiers killed in action at the stockade or wounded either mortally or severely. There is no mention, or even indication, of men who might have suffered light wounds.

The circumstances of the battle indicate that there just must have been more soldiers injured than reported. The melee that occurred when the Redcoats first entered the stockade was a very close, personal, and brutal affair. Simple common sense dictates that the insurgents would have struck out and injured some of the Redcoats. Many of those injuries might not have been reported.

Soldiers throughout the ages have been reluctant to acknowledge wounds or injuries they have suffered, particularly what might be considered slight wounds. They might feel that they can deal with such matters themselves, or that to make a fuss about a light wound when others have suffered much worse was somehow demeaning of both themselves and those more seriously wounded. Anecdotal evidence from modern wars indicates that soldiers who have been lightly wounded are more likely to self-administer first aid than report the wound officially. Veterans of the Second World War, Korea, Indo-China and Iraq all tell stories of ignoring light wounds. Soldiers often claimed to have 'got nicked' or taken 'a little shrapnel'. They might also ignore wounds that would technically have entitled them to an award, or have to be ordered to have light wounds treated. In all cases, the dismissive response to light wounds by individual soldiers is consistent.[61] The men of the 12th and 40th would have been no different.

Even quite serious injuries might have been omitted from the official records for Eureka, especially if the soldier presented for treatment some time after the battle. Adeliza, the daughter of Mary Faulds, recalled that at the 50th anniversary of Eureka an old man who claimed to have been a soldier there pointed to an eye that had been poked out, and said that he lost it in the battle for the stockade. No such injury was recorded on the official casualty returns for Eureka. The claim might have been nonsense, or it might not have been.[62]

How many soldiers at Eureka were wounded lightly, or even quite seriously if the eye story can be believed, can only be guessed. Luckily, at least one close-to-contemporary source exists that can help to make an informed estimate. The casualty returns for the British Legion in Spain between 14 May 1836 and 10 May 1837 list 1351 admissions to the hospital, of which 585 were described as '[g]eneral wounds, slight'.

The number of these slightly wounded men who were subsequently discharged for duty was 548, with 34 being transferred and three dying.[63] These figures indicate that the number of slight wounds suffered by the British Legion during that period was about 43 percent of the total casualties, a substantial fraction. It can be assumed that the soldiers at Eureka also suffered their share of 'general wounds, slight' that were never recorded in the official casualty returns.

One reason that many wounds inflicted on the soldiers and police might have been less than traumatic was the nature of the weapons the insurgents used. The double-barrelled shotguns and fowling pieces carried by many were only effective at the closest ranges. The wounds inflicted by them would have depended on the distance to the target, the weight of shot fired, and the steadiness of the person firing.[64]

The small calibre handguns carried by many insurgents might also not inflict serious injury. Pocket watches, buckles, coins, small books and even thick coats could deflect or retard balls from such weapons.[65] Diggers fists, boots and teeth would inflict painful, but not life threatening, injuries. Improvised clubs, shovels, pike blades and knives all played their part in the melee that engulfed the stockade when the soldiers stormed in, but it would have been a matter of luck if these weapons caused serious or mortal wounds.

There were claims of quite heavy losses being suffered by the soldiers, with accounts mentioning from ten to 18 being killed, and up to 20 wounded.[66] Miner James Smith, relating a third hand story, told how an American insurgent killed seven soldiers before being himself killed, and that the military losses were heavy, but concealed by being recorded privately and never released to the public.[67]

Apart from providing fodder for conspiracy theorists, these claims cannot be taken seriously. It would be most unlikely that deaths could be hidden so completely that they would remain so to the present day. Despite the questionable nature of the 40th's record keeping at the time of Eureka, details of pay and pensions were another matter, and careful records of these were maintained. The number of deaths officially reported for the 40th in 1854 showed no increase from the previous two years. This could seem strange, but was not necessarily so, as in 1855 the death rate was less than half that of 1854, showing there could be considerable variation from year to year.[68]

A letter to the *Argus* from a miner, discussing the military casualties at Eureka, made the point that reports 'are so contradictory, it is impossible to say; but I believe that two are killed beyond doubt, and some half dozen or half score wounded'.[69] Another claim, many years after the event, was that there were 29 total military casualties, eleven more than otherwise claimed. No evidence was given, other than recollections of anecdotal accounts heard by the author when a child, from people who lived on the diggings at the time of Eureka. It is interesting, however, that while the wounds mentioned in this account were all described as serious, the figure of 29 is close to what might have been the actual casualties, if slight wounds not reported by the soldiers are taken into account.[70] In one history, a claim was made that ten to twelve soldiers died, but that their deaths were concealed so as not to give solace to the disaffected.[71] Again, no evidence was provided.

The special Ballarat correspondent for the *Argus* saw seven insurgents and six soldiers in a hospital after the battle, including the injured drummer boy from 28 November.[72] This was only half the number of soldiers officially listed as seriously wounded. All or some of the remainder appear to have been transferred to Geelong, Alpheus Boynton, a carter, reporting seeing coffins and injured men on the road from Ballarat to Geelong after Eureka.[73] Numerous soldiers of the 12th spent from one to 27 days in hospital following the battle. This might have been due to an outbreak of dysentery that occurred in the government camp, or, at least in part, the result of wounds, serious or slight, suffered during the battle.[74]

There was a hospital at the government camp and two private tent hospitals on the Ballarat diggings, one at Bakery Hill and one at Red Hill. Doctor Francis Carr and another doctor named Clendinning entered the stockade immediately after it fell to the army and police. They at once moved several wounded to the Red Hill hospital for surgery. A letter from D.J. Williams, the District Surgeon, referred to the government hospital being unable to take any wounded.

The Red Hill hospital received the casualties mentioned by the *Argus* correspondent. The last of the wounded prisoners was discharged on 10 January 1855, five and half weeks after Eureka. Carr later claimed he spent three months working every day to care for the wounded, presumably military casualties.[75]

One indicator of the serious nature of the battle is the percentage of the total force engaged who became casualties. Victors in battles fought in open country during the Napoleonic Wars, an era of similar military technology to Eureka, normally took about nine percent losses.[76] There were 276 soldiers and police at Eureka, and of these the 176 infantry bore the brunt of the fighting. The 16 officially acknowledged casualties at Eureka, which do not include McIvor, equal nine percent of the infantry force present. The addition of Brien, and possibly Hall and Hammond, to this number would increase the casualty rate to eleven percent. These losses indicate quite clearly the significance of the struggle that morning. Experience of battle through the ages makes it certain that more Redcoats were injured at Eureka than was officially acknowledged.

The exact number of insurgents who died because of Eureka will also never be known. A clue can be gleaned from the number of shots fired by the soldiers. We have already established the probability that they fired hundreds of shots at the stockade. Based on the likely number of hits from musketry, and the casualties from bayonets, the losses suffered by the insurgents at Eureka must have been many more than the 22 on Lalor's list, or seen by Mrs Shanahan inside the stockade following the battle.[77]

Another indicator that many insurgents died was given by the *Argus*, which reported that the 'wounded are dying fast; nearly all the wounds got in the affair are proving mortal'.[78] In the letter from a miner to the *Argus* referred to earlier, in which he attempted to determine how many soldiers fell, he also wrote that as 'far as I can learn, the number killed on the part of the mob was forty; the number wounded I could not ascertain'.[79]

In his report on the battle, Thomas recorded that at least 30 insurgents were killed outright.[80] Of fundamental importance in determining that their losses were greater than admitted are the 21 unidentified bodies that were buried after Eureka. The addition of these unknowns to Lalor's list gives a total of 43 killed. Pierson thought that the final total of dead was about 60, with many lying at the bottom of mine shafts they had jumped or fallen into, or going into the bush to die.[81] Miner William Adams also believed that insurgents had jumped down mine shafts to avoid capture.[82] Some probably died that way.

A principal reason the records of those who died might not be accurate is that the official death notices were not recorded until 20 June 1855, six and a half months after Eureka, leaving ample opportunity for omission by mistake or intent.[83] Many insurgents would have fallen during what Lynch called the fierce 'saturnalia of carnage' that engulfed the stockade, and estimates such as Pierson's 60 dead may be close to the actual losses.[84] It remains a sad irony that, because the names of many insurgents who died defending the stockade were never recorded, or their deaths listed on any monument or memorial, they have simply vanished from history.

Following the fall of the stockade, triumphant mounted troopers boasted to those insurgents they were escorting that 'Joe is dead now'.[85] Their bravado was that of men who had come through an unexpectedly bitter and hard fought engagement. In their exhilaration of victory, they naturally felt empowered to taunt their defeated foes.

Eureka was a battle that erupted unexpectedly and was fought out with unanticipated ferocity, resulting in bloodshed and carnage. This was the stark reality, and only by appreciating this can we understand what happened there.

Chapter 10
'THE BONNY BRIGHT FLAG'

When Burnette fired his rifle and hit a soldier in the early morning light, he could hardly have imagined that his single shot sparked an event that, more than 150 years later, still inspires passionate debate and reflection. For around 20 minutes after his shot, men fired, lunged and stabbed musket, double-barrelled shotgun, carbine, pistol, revolver, bayonet and pike over little more than a crowded acre (0.4 hectare) of fenced-in mud and clay.

Eureka was not a simple riot, a description deliberately used at the time and since to demean the insurgents' cause and detract from the significance of the event. It was also not an unprovoked, wild rampage by uniquely brutish soldiers and cowardly, murderous police, a depiction favoured by those who sought to portray the government of the day in as harsh a light as possible. While it lasted, the battle for the stockade was a military engagement as fierce and cruel as any. When the shambles abated and the passions cooled, scores lay dead and wounded. They were casualties of an act of war, nothing more, nothing less. Recognition of this grim fact sets the conflict at Eureka apart from every other example of communal dissent in Australia's history.

The battle was not as hopelessly one sided as myth insists. The insurgents were not the helpless or poorly armed innocents that popular legend insists. They were not taken by surprise, nor were they asleep or hopelessly drunk. They were not greedy, self-centred, tax evaders. The vast majority of them were not anarchists, socialists, or revolutionaries.

Nor were the majority foreigners, the 'mongrel crew of German, Italian and Negro rebels' decried by one newspaper at the time.[1]

Despite the colonial government's grave fears to the contrary, most insurgents at Eureka were not even fighting for radical changes to the established political order. They simply wanted to be allowed to enjoy a vestige of individual dignity that they felt law, tradition, and custom, entitled them to as Britons. When that dignity was emphatically denied to them at the point of the bayonet, they rebelled, and an army of civilians, the Gentlemen Soldiers, came into being. Consumed with righteous fury, they naïvely imagined that by parading and posturing as soldiers, they could force the goldfield authorities to capitulate, or at least get the government to concede their case. History records that in the first hope they were tragically mistaken, but they were overwhelmingly successful in the second.

Eureka Veterans. In 1904 the proud veteran miners who fought at Eureka gathered for the 50th anniversary commemoration of the battle. (Leader 1904)

The Eureka miners' cause did nurture within it the seeds of greater things. The Reform League's Charter of 11 November 1854 set out an agenda for radical political change that challenged the very foundations of colonial authority. When an armed insurgency developed, which appeared at first glance to be linked to those claims, the government of Victoria was compelled to take active measures to crush what they saw as a revolution in the making. It was fear of the potential for violent, radical, change that incited men such as Pasley, who had extraordinary influence at Ballarat, to insist in increasingly bellicose terms on physically eliminating the threat before it brought what he deemed to be ruin to the colony.

We have seen that the legend of Eureka developed very quickly to exclude deliberately groups that, for whatever reason, fell from favour at the time. The most obvious and significant of these were the Americans and the police. The very significant role the Americans played in the defence of the stockade, while briefly alluded to by a few writers in the past, has never been fully acknowledged until now. The decisive tactical contribution made by the mounted police, whose role at Eureka has previously been interpreted solely as that of cowards, thugs and murderers, was also studiously ignored.

While not the genesis of democracy in Australia, as passionately claimed by some, Eureka surely played the role of its most adept midwife. The bloodshed there indelibly stamped the character of the inclusive Australian democracy that developed in following decades. The armed rebellion at Eureka and the mass public protests that erupted in Melbourne, supporting the Eureka accused during their trials for treason, emphasised the dangers for any Australian government that spurned the aspirations of those it governed. Eureka was very much, as Clive Turnbull once observed:

> an affirmation of Australian nationhood and Australian democracy
> … both a warning to would be tyrants and an inspiration to those
> who, like Ross and Thonen, are willing to give even their lives in
> love of liberty and their fellow men.[2]

The subsequent influence of this Eureka spirit on the Australian political experience cannot be denied. Political and industrial activists of all shades and opinions have frequently evoked the spirit of resistance to what they consider obnoxious authority, embodied by Eureka, as their inspiration when making one partisan point or another.

Even the symbol of Eureka has left an indelible mark on Australian society. Many in the community who, for whatever reason, have felt their collective rights to be threatened, or simply wished to announce their solidarity with their peers, have adopted Eureka's Southern Cross flag. The Eureka ensign has become so associated with the culture of protest that it has become something of a symbol for the radical fringes of politics. This has contributed significantly to its marginalisation, and exclusion from the mainstream of political consciousness. The Southern Cross flag has become, in the eyes of many, a somewhat disreputable object, rather than the icon of truly national significance that it is. That this was allowed to happen is a tragedy of monumental proportions.

The Eureka Flag. The original flag flown by the insurgents over the Eureka Stockade is displayed in all its glory at the Ballarat Fine Art Gallery. (Collection: Ballarat Fine Art Gallery)

The actual starry banner still exists, displayed in its own hall at the Ballarat Fine Arts Museum. The 'Bonny bright flag of the Southern Cross'[3] is a direct physical link to an event and an era that contributed fundamentally to our national heritage, and is an historical artefact of immense significance. Given what we now know of the true nature of the battle for the Eureka Stockade, surely the time has come for the flag that flew above the brave insurgents that fateful morning to be granted the official national acknowledgement and prominence it deserves.

In this narrative, we have journeyed back in time, and for a while climbed into the skin of those who were present at Eureka. We have understood what it was like to have been a participant on both sides of the conflict, and by doing so, determined what motivated such a decisive and deadly clash. We have seen what it was like to have been an insurgent crouched behind the slab palisade, trading shot for shot with the oncoming soldiers, and experiencing the range of emotions, ranging from dour, stubborn, refusal to give ground, to sheer terror when everything collapsed in ruin.

We have also seen what it was like to have been a soldier, advancing cautiously against an unexpectedly determined and effective resistance, and then surging forward, blood up and bayonet to the fore. In the same way, we have learned what is was like to have been a policeman during the battle, and realised the decisive role they played in breaking the insurgents' resistance on the day.

The battle for the Eureka Stockade was a tragedy of the classic sense. Two groups of protagonists, from the same community, came to a point where their equally passionate, but opposed, senses of duty, tradition and propriety clashed head on. The result was the bloody conflagration at Eureka.

It has been said that history is a myth that men have agreed to believe.[4] It is in this manner that the story of the Eureka Stockade has become part of our collective memory. Like all collective memories, our understanding of what occurred during the battle is anti-historical, simplified, and viewed from a single perspective, notably that of an aggrieved victim. This memory has been stubbornly uninterested in any ambiguities that challenge it.[5] Until now, such has been the fate of the narrative of Eureka. If nothing else, this book has laid to rest the fallacies that have obscured and complicated our understanding of what happened there.

We surely owe at least that much to the memory of those who died at Eureka on that Sunday morning in December 1854.

Appendix 1
THE CHARACTERISTICS OF FIREARMS USED AT EUREKA

Revolvers

The versions of the Colt revolver available to the insurgent diggers of 1854 were the .3[1] calibre 1849 model 'Pocket' Colt, the .36 calibre 1851 model 'Navy' Colt and the massive .44 calibre 1848 model 'Walker' Colt1. McGill, leader of the Independent California Rangers Revolver Brigade at Eureka, carried a Walker Colt.

The Colt revolver, like all percussion cap weapons, could misfire at times and, in common with all such weapons of the period, was not particularly accurate unless fired at very close range. Another idiosyncrasy of the Colt was that all of its chambers might sometimes discharge at once, a most disconcerting event for both the firer and target. Despite such shortcomings, its rate of fire, sturdy construction and general reliability made it a weapon not to be despised, and one that was certainly used by a significant number of Eureka's defenders. A good example of the limited accuracy of such revolvers was the 1852 duel fought in California between US Senator David Broderick, a noted good shot with the pistol, and Caleb Smith. Both men used Navy Colt .36 calibre revolvers. They fired at each other at close range, presumably at the 20 paces or so that seemed to be the standard for such affairs. Each man emptied his revolver at his opponent, with the only effect that Broderick's life was saved when a ball from Smith's revolver was deflected after it struck his pocket watch.[2] Another duel in San Francisco in 1858, between George Pendleton Johnson and William I. Ferguson, resulted in the two men exchanging three shots at each other at a distance of 20 paces. Johnson hit Ferguson only when the range was shortened to 20 feet (six metres), but then only in the thigh.[3]

Another revolver very common on the goldfields was the 'Pepperbox'. This had six pre-loaded barrels arranged in a circular pattern around a central pin. When its hammer was cocked, either physically, or by squeezing the trigger, a barrel rotated into position for firing. Coming in several variations, Pepperbox revolvers first appeared in the late 1830s and continued to be manufactured through to the 1870s, even when the more modern Colt revolvers superseded their design. While a robust weapon, the Pepperbox was woefully inaccurate. American author and raconteur Mark Twain recounted how a Pepperbox had been fired at a playing card nailed to a tree, and, missing the card, managed to strike a mule some 30 yards (27 metres) to the left.[4] In another reminiscence that highlighted the unreliability of the Pepperbox, Twain described how two men trying to kill their uncle with a Pepperbox had snapped the hammer of the revolver repeatedly in his face without it discharging.[5] William Gay, the son of a miner at Ballarat during the Eureka period, explained that the Pepperbox, which he misnamed as a 'paper box', needed 'a good shot to hit a haystack with them at 20 yards [18 metres]'.[6] Like the Colt, the Pepperbox also sometimes unexpectedly fired all its barrels at once.[7] Pepperboxes were in widespread use among gold miners, and they would have been present in the stockade.[8]

Shotguns and Rifles

The most common long-arm carried by the insurgents at Eureka was the muzzle-loading shotgun, called the 'gun' or fowling piece in contemporary accounts. These came in either double-barrelled or single-barrelled versions. The most common ammunition for them was loose buckshot, not heavy bullets, as their thin barrels could not accept the heavy loads of powder required for the latter. Buckshot was of limited military use at ranges over 60 yards (55 metres). In a test conducted for the TV series Battlefield Detectives, a load of buckshot was fired from a smooth bore firearm at a human torso-sized flat target at this range. Less than half of the load of buckshot balls hit the target, demonstrating the poor ballistic characteristics of such ammunition. During the test, the comment was made that buckshot fired at a greater range would have lost much of its energy, and be unlikely to inflict serious injuries. Some of the insurgents, such as Burnette, carried rifles. Unlike shotguns, the rifle fired a single solid bullet, and while smooth bore weapons were notoriously inaccurate, the rifle's accuracy was noticeably superior. Although the rifles used at Eureka would have been muzzle-loading, and needed special skills to use efficiently, they could be formidable in the hands of a marksman.

Military Muskets

The Lovell's[9] 1842 Pattern smooth bore muzzle-loading musket carried by the soldiers at Eureka was essentially an upgraded version of the tried and true Brown Bess musket with which the British army had been equipped, in one form or another, since the early years of the eighteenth century. The 55-inch (1.4 metre) long 1842 Pattern musket had a heavy wooden butt and fore stock, and weighed ten pounds (4.5 kilograms). Like all smooth bores, its accuracy left much to be desired. While the ball could travel up to 700 yards (over 600 metres), a hit at that range would be entirely due to uncannily bad luck for the target, and be essentially harmless. The old adage about not firing until you could see the whites of their eyes had real meaning for men armed with smooth bore weapons. The appallingly bad accuracy of muskets had a decisive influence on how and where the fight for the stockade occurred.

The only real improvement in the performance of the 1842 Pattern musket over its ancestors was the system of percussion cap ignition it used to detonate the black-powder charge. The percussion cap proved much more reliable than the older flintlock system, cutting down significantly on the prevalence of misfires that had been a constant source of frustration. The percussion ignition system was simple. A small copper cap, lined on its interior with a thin film of a fulminate, usually of mercury, was placed over an open nipple that led directly into the chamber of the weapon. When the trigger was pulled, it released the musket's hammer, which struck the cap. The shock ignited the fulminate and caused a flash that travelled down the tube of the nipple to the black-powder rammed into the chamber, and started an explosion, propelling the musket ball out of the weapon. It was a simple and effective system.[10] A small pouch on the soldier's cross-belts, or waist-belt if he was so equipped, held his percussion caps. To fire a muzzle-loading weapon, the soldier first extracted a cartridge from the cartridge pouch that hung from his cross-belt and rested on his right hip. The pouch normally held 50 or 60 cartridges, arranged in two layers. Because of the extreme awkwardness of extracting the cartridges in the second layer, experienced soldiers removed some and tucked them into the waistband of their trousers before a battle, for easier access. The cartridge was a paper tube holding a measure of black-powder and a lead bullet. Wax or another grease was used to seal and waterproof it. When loading his musket, the soldier took the cartridge between his teeth and tore off the opposite end to the ball. Then, with the butt of the musket resting firmly on the ground, he poured the black-powder down the barrel. The bullet was then inserted into the muzzle, with the paper on top to act as wadding. Both ball and paper were rammed down the barrel using a metal ramrod, which was housed in a groove beneath the musket's fore stock when not being used. The musket was then lifted, and its hammer moved back to the half-cock position. This locked the hammer, and was a necessary precaution to prevent accidental release while the percussion cap was being placed, which would cause a potentially catastrophic premature ignition of the charge. Once the percussion cap had been placed on the nipple and the musket lifted into the firing position, the hammer was pulled back to its fully cocked position. The soldier took aim, pulled the trigger, releasing the hammer that struck the cap, and the musket fired with a heavy dull thump, flash of flame, eruption of thick white smoke and mule-like kick. An experienced soldier then pulled the hammer back to half-cock, flicked away the expended cap, placed his lips over the muzzle and blew hard, forcing a thin plume of smoke from the nipple tube. This extinguished any lingering embers in the chamber which, if ignored, could result in a dangerous flash ignition when he reloaded the musket. The process was then repeated. A trained soldier could load and fire a musket three times in one minute. If more rapid fire was desired, he could omit the paper wadding or ramrod and simply pour the powder and drop the ball down the barrel. The charge and ball were then seated into place by tapping the butt of the musket firmly on the ground several times. This method greatly reduced the accuracy of the already inaccurate musket, but increased the rate of fire to four or five rounds a minute. Such rapid fire had its drawbacks, as after firing 20 or 30 rounds, the musket became too hot to handle.

The Need To Aim Low

A peculiarity of smooth bore weapons of the era was that the trajectory of the balls tended to rise when they left the muzzle. This presented a problem for men who were firing at targets on lower ground. To compensate for this, the musket needed to be aimed very low. Firers, especially inexperienced ones in the heat of battle, generally failed to comply with this necessity, and most did not adjust their aim. That, combined with the bad habit of lowering the butt of the weapon from the shoulder to avoid the savage recoil of black powder long-arms, resulted in many shots flying harmlessly high. J.W Baldwin, a private in the 9th Regiment during the Sikh War of 1845-46, related how during one engagement the fire of the regiment was directed very low, and did great execution, while that of the enemy passed over their head 'as thick as hailstones'.[11] The tendency for shots to fly high was especially evident at close ranges like those at Eureka.[12] Even in non-military confrontations, the phenomenon of firing above a target when shooting from higher ground occurred. A duel fought between two shotgun-armed Californians during the early 1850s resulted in one of the duellists putting his shot just above the head of his opponent whose aim, being truer, killed his adversary. The consensus of those who witnessed the duel was that the unfortunate victim was at a disadvantage because he stood on slightly higher ground, and in the heat of the moment did not depress his weapon enough, thus missing his target.[13] Thomas would have been very aware of the idiosyncrasies of the firearms of the day. While uncertain exactly what weapons the Eureka insurgents were armed with, he could anticipate that their aim would probably be poor, as a consequence of forcing them to shoot down a slope.

The Sound made by Bullets of the Era

One idiosyncrasy of black-powder weapons is the sound their bullets make as they cut through the air. In *The Face of Battle*, Keegan recounted how soldiers who fought at Waterloo, where the same type of musket balls fired at Eureka were used, reported that they made sounds like 'whizzing', and 'whistling'. This was the case at Eureka. R. Lorimer, who was boarding in a restaurant near the stockade, recalled that the 'bullets were whizzing about like mosquitoes'.[14] Personal experience of firing a black-powder Enfield rifled-musket, and having a round from a black-powder Colt revolver fly close by the author's ear has confirmed these reports.

The Effectiveness of Military Musketry during the Black-Powder Era

At the time of writing in the early twenty-first century, many assume that when using a military firearm, what one aims at will probably be hit. This is not correct now, and certainly would have been a ludicrous proposition in 1854. George Nafzinger described the accuracy of smooth bore muskets in a well-argued article titled *The Power of Musketry in the Napoleonic Wars*. The ballistic characteristics of such muskets were essentially the same as the muskets used by the soldiers at Eureka. Using empirical evidence to arrive at the percentage of shots that could be expected to hit a given target, Nafzinger concluded that when veteran troops were firing, 40 percent of shots could be expected to hit a target at 150 yards (135 metres), falling to 25 percent with novice troops. Such a rate of hits might seem adequate, but these figures are for targets in good light, with no cover, and by troops who are themselves not being shot at. Brent Nosworthy in *Battle Tactics of Napoleon and His Enemies* was much more conservative in his estimates of the effectiveness of smooth bore musketry. He used historical evidence from studies conducted during the nineteenth century, which compared the number of rounds actually fired during battles to the number of casualties caused, to conclude that the actual rate of hitting a target varied between 0.30 percent and 2 percent. In *Firepower – Weapons effectiveness on the Battlefield, 1630-1850*, B.P. Hughes explained the idiosyncrasies to be expected from the smooth bore musket in battle. Despite theoretical hit rates of 75 percent at 100 yards (90 metres), usually against solid targets in optimum conditions, by a careful analysis of statistics and consideration of what he deemed 'Degrading Factors' related to the 'Inefficiencies of the Battlefield', Hughes arrived at an actual number of balls that 'reached their mark' of less than 50 percent at that range, with only marginal improvement at closer ranges. From those, he further refined his figures to arrive at a figure of 5 percent actually being effective. This might have been slightly exceeded by good quality troops. Degrading factors, all of which affected a weapon's effectiveness, included an animate target, technical failures, human error, the ground, ammunition supply and smoke. The insurgents at Eureka were in cover, in poor light and wreathed in smoke, making it easy to see how Hughes' degrading factors would have applied. Shanahan's nine casualties from probably more than 100 musket balls fired in the soldiers' first volley sounds about right in the circumstances. It was seldom the sheer weight of firepower alone that decided matters in the smooth bore era, especially against an enemy behind adequate cover, as were the defenders of the stockade. Failure of morale and the vicissitudes of battle more often than not turned the tables on one side or the other, and such eventualities occurred during the battle for the Eureka Stockade.

Wounds Caused By Black-Powder Weapons

The wounds caused by black-powder weapons are very different in character to those caused by modern firearms. The soft lead balls could produce horrific injuries but, if vital organs were missed, a wound did not necessarily cause immediate disablement of the victim.[15] There is a ndistinct difference between the effects of the smooth bore black-powder weapons carried at Eureka and a modern firearm. A modern military rifle, such as the M16-A2, fires a round that travels at 2800 feet (850 metres) per second.[16] When this projectile strikes a target at short range, it retains a great deal of energy. When that target is soft, such as the human body, this causes massive trauma. As the round penetrates, the unexpended energy is released as shock waves that travel at about 4800 feet (1460 metres) per second, the same speed as sound travels through water. This causes an effect known as cavitation, where a cavity forms immediately behind the passage of the bullet. Tests using gelatin blocks, which have the same consistency as human flesh, indicate that this cavity can be as much as ten times the size of the calibre of the round. The shock waves that accompany cavitation pulp tissue and smash bone within the radius of the effect, causing serious internal damage. When a modern round has passed through or stops inside the body, the entry passage collapses to a narrow pathway that belies the damage. The musket and pistol balls fired at Eureka behaved in an entirely different manner. Because they travelled at the relatively slow velocity of 600 feet (180 metres) per second,[17] the lead ball tended to burrow into flesh, leaving a passage of uniform diameter, the same as the calibre of the ball. Only the tissue in the immediate vicinity of this wound passage was damaged. There was no cavitation effect or collateral damage to surrounding tissue. The result was that vital organs or bones that were not hit were not damaged and a victim could, if fortunate, remain able to move and fight.

Appendix 2
LEGAL ASPECTS OF THE SUPPRESSION OF RIOT, INSURRECTION AND REBELLION IN 1854

A great deal was made at the time of Eureka, and later, of the allegedly unconstitutional, and therefore illegal, nature of the government's action in using armed force against the insurgents. Lalor made this accusation directly[1]. The words '[r]esisting the unconstitutional proceedings of the Victorian Government' were literally engraved into posterity on the monument over the grave of the fallen Eureka insurgents.[2] At a rally held on St Paul's Church reserve in Melbourne on 6 December 1854, a resolution was adopted, stating that the 'constitutional agitation at Ballarat has assumed the present unconstitutional form in consequence of the coercion of military force'.[3] The basis for this claim was that the soldiers fired first, without the reading of the Riot Act, and that every action taken subsequently by the army and police to suppress the insurgents by force of arms was therefore unconstitutional and illegal. These assumptions are quite wrong on both counts.

The notion that the soldiers fired first is simply incorrect. Also, the law then regarding suppression by the authorities of riots and illegal gatherings was quite clear that the Riot Act need not be read.

In charging the Grand Jury at the 1832 Bristol Special Commission following the Reform Bill Riots of 1831, Justice Tindal stated that the military authorities were bound to do their utmost, on their own responsibility if necessary, to put down any situation of a riotous and tumultuous nature. By referring to them operating on their own responsibility, Tindal conceded that in certain circumstances the military authorities could act without obtaining the consent of the civil authorities. By using the phrase 'bound to', the mandatory nature of the military response was made clear. In a footnote, he stated that a riot or illegal meeting was no less riotous or illegal because the Riot Act had not been read, and all parties that did not disperse were still committing a capital offence.[4]

In 1838 the then British Attorney General, Sir John Campbell, and the Solicitor General, Sir R. M. Rolfe, ruled that a governor of a district could, without declaration of martial law, put down any insurrection or rebellion by force of arms, and lawfully put to death any person engaged in resisting such actions.[5] In neither of the cases ruled on by Tindal or Campbell and Rolfe was the presence of magistrates deemed essential, or the reading of the Riot Act considered obligatory. This interpretation of the law related to the suppression of riots and disorder remained current until 1928.

This was the law that governed the reaction of the colonial government and its military and police forces to the events at Eureka. Subsequent claims that the government acted in an unconstitutional and thus illegal manner are flawed. In commenting on Hotham's actions, Blainey emphasised that in 1854, every governor in the western world would have acted similarly, if faced with such a rebellion.[6] It would have been a particularly impotent or broad-minded government of the era that would not have put down by force the armed insurrection that occurred at Ballarat. Unfortunately for the insurgents at Eureka, the Victorian colonial government in December 1854 could not be accused of being either impotent or broad-minded.

Appendix 3
DRAMATIS PERSONAE

This list makes no pretence to be a comprehensive catalogue of everyone who was at the stockade or involved in the events at Eureka. Those named here are mentioned in the text of this book, and were involved in some aspect of the events that led to the battle for the stockade, the battle, and its aftermath. Parentheses mark where names are spelt in different ways in more than one source. Wherever possible the nationality of the insurgents is mentioned.

Government and Administration

A'Beckett, William	chief justice of Victoria, chief prosecuting counsel at the Eureka Treason Trials
Amos, Gilbert	Gold Commissioner
Barry, Redmond	Judge presiding over the Eureka Treason Trials
Foster, John	Chief Colonial Secretary of Victoria
Hackett, Charles	Magistrate
Hotham, Sir Charles	Governor of Victoria
Huyghue, Samuel	government employee - clerk
Rede, Robert	Resident Gold Fields Commissioner, Ballarat
Webster, George	Magistrate

The Soldiers

Adams, Lieutenant	40th Regiment
Atkinson, Captain Arthur	40th Regiment, commanded the government camp at the time of the battle
Bodely, Private Thomas	40th Regiment

Boyle, Private Felix	12th Regiment, died of wounds
'Brennan' (Brewer), Sergeant	40th Regiment, said to have encouraged insurgents to leave the stockade and waste ammunition.
Brien, Private Dennis	40th Regiment, killed
Egan, Drummer Boy John	12th Regiment, wounded 28th November
Gore, Private James	40th Regiment
Hall, Lieutenant Charles	40th Regiment – mounted
Hammond, Private James	died on the road to Geelong following Eureka
Harris, Sergeant Edward	40th Regiment
Hegarty, Sergeant Daniel	40th Regiment
Juniper, Private William	40th Regiment, wounded
Littlehales (Little), Captain George	12th Regiment, died of disease following Eureka
Lough, Private James	40th Regiment
Lynott, Private Patrick	40th Regiment
Neill, Private John	40th Regiment
O'Keefe, Private Patrick	40th Regiment
Pasley, Captain Charles	Royal Engineers, ADC to Thomas
Paul, Lieutenant William	12th Regiment, wounded
Queade, Captain William	12th Regiment
Revell, Private William	40th Regiment – mounted
Richards, Lieutenant Thomas	40th Regiment
Richardson, Corporal William	40th Regiment – mounted
Riley, Sergeant Patrick	40th Regiment – mounted
Roney, Private Michael	40th Regiment, killed.
Shanahan	soldier
Sullivan, Private John	40th Regiment
Thomas, Captain John Wellesley	40th Regiment, commanded the attack
Wall, Private John	12th Regiment, died of wounds
Wise, Captain Henry Charles	40th Regiment, died of wounds
Wood, Private George	12th Regiment

The Police

Badcock, Constable John	foot policeman
Carter, Sub-Inspector Charles Jeffries	commanded the Foot Police

Chomley, Sub-Inspector Hussey	mounted policeman
Culkin, Trooper John	mounted policeman
Foster, Inspector Henry	mounted policeman
Furnell, Sub-Inspector Samuel	mounted policeman, commanded the right wing of Thomas's force.
Gillman, Sergeant	mounted policeman
Glover, Constable Joseph	policeman
Goodenough, Henry	police spy
King, Constable Hugh	foot policeman
King, Constable John	foot policeman, took down the Southern Cross flag
Kossak, Sub-Inspector Ladislaus	mounted policeman
Langley, Sub-Inspector James	mounted policeman
Lawler, Sergeant Major Michael	policeman
McIvor (McIver), Constable	mounted policeman, wounded
Milne, Constable Thomas	mounted policeman
Peters, Andrew	police spy
Thompson, Constable William	mounted policeman
Viret, Sergeant Edward	mounted policeman
Ximenes, Sub-Inspector Maurice	mounted policeman

The Insurgents

'American Captain' (per Carboni)	directed the defence of the stockade
'Blackjack'	American, 'killed at Eureka'
'Oravalano'	Italian insurgent
Allan, Richard	Scottish insurgent
Ashburner, James	English insurgent
Bailey, A.A.	American, possibly killed at Eureka
Beattie, James	English insurgent
Black, George	member of the insurgent council
Brown, James	American insurgent
Burnette, Robert	American insurgent, California Ranger
Callinan, Patrick	Irish insurgent
Campbell, James McFie	West Indian insurgent
Canny, Michael	Irish insurgent, wounded
Canny, Patrick	Irish insurgent, wounded
Carboni, Raffelo	Italian insurgent, witness to the battle
Curtain, Patrick	Irish insurgent, captain of the pikemen

De Longville, Henry	insurgent
Diamond, Martin	insurgent, killed
Dignum, Thomas	Australian insurgent
Esmond, James	Irish insurgent – discoverer of gold
Ferguson, Charles Derius	American insurgent, California Ranger
Flynn, Ned	insurgent, killed
Gettings (Gittens), Patsy (Patrick)	Irish insurgent, killed
Hafele, Johann	German insurgent, killed
Hall, Walter	American insurgent
Hanrahan, Michael	Irish insurgent, captain of pikemen
Hartley, George	American insurgent
Hull, James	American insurgent
Hynes (Hines), John	Irish insurgent, killed
Joseph, John	American insurgent
Lalor, Peter	Irish insurgent, commander in chief of the insurgents, wounded
Lynch, John	Irish insurgent
Madden, John	insurgent
Madocks, Alfred	English insurgent
Manning, John	'Irish' insurgent, journalist
McGill, James	American insurgent, commander of the California Rangers, second in command from Saturday night
Melody, Bill	American insurgent, California Ranger
Miller, Montague	Australian insurgent, pikeman, wounded
Molloy, Patrick	insurgent
Moore, Teddy	insurgent, killed
Murphy, Patrick	insurgent
Perry, Henry	insurgent
Nelson (Nealson)	American insurgent, commander of the First (Californian) Rifles
O'Neill, Thomas	Irish insurgent, pikeman, killed
Phelan, John	insurgent
Read, Henry	insurgent
Robertson, John	Scottish insurgent, killed
Rodan, John	insurgent
Romeo, John Francis	insurgent
Ross, Charles (Henry)	Canadian insurgent, commander of a 'rifle' company, killed

Shanahan, Edward	insurgent, shopkeeper inside the stockade
Sutherland, Henry	insurgent
Thonen, Edmund	German insurgent, commander of a 'rifle' company, killed
Tuohy, Michael	Irish insurgent
Vern, Frederick	German insurgent, second in command of the insurgents for a time

Observers and Citizens

Allen, Thomas	storekeeper, Waterloo veteran, critic of the insurgents
Amies, Elizabeth	18 months old, might have been in the stockade during the attack
Arnold, A.W.	witness to the storming of the stockade.
Aspinall, Butler Cole	John Joseph's defence counsel
Bentley, James	proprietor, Eureka Hotel
Bourke, W	witness to the storming of the stockade
Brandt, Johan	proprietor, Prince Albert Hotel
Budden, Thomas	Canadian, friend of Charles Ross
Carr, Francis	doctor at Ballarat
Clendenning, George	doctor at Ballarat
Dynan, Michael	inside the stockade
Faulds, Mary	inside the stockade
Faulds, Matthew	inside the stockade
Gregorious, Johannes	servant to Catholic priests at Ballarat
Grieg, Agnes	witness to the battle
Humphris, Mary Jane	witness to the battle
Lazarus, Samuel	shopkeeper, witness to the aftermath of the battle
Nicholls, C.F.	visited the stockade prior to the attack
O'Brien, John	claimed to be at the stockade
Parker, 'Mrs'	witnessed the battle
Pierson, Thomas	American miner, lived near Bakery Hill
Shanahan, 'Mrs'	inside the stockade
Smyth, Patrick	Roman Catholic priest at Ballarat
Weisenhaven, Carl	worked at the Albert Hotel

SELECT BIBLIOGRAPHY

ARCHIVES AND UNPUBLISHED PAPERS

Victorian Archives

AJCP M2579, 40th (2nd Somerset Regiment) Service in Australia

AJCP M2580, 40th (2nd Somerset Regiment) Service in Australia and Action at the Eureka Stockade

1855 Victoria, State Trials, Queen v. Hayes

1855 Victoria, State Trials, Queen v. Joseph

Eureka Documents 1, three dispatches from Sir Charles Hotham, Public Records Office, Melbourne, no date

Map: Geological Survey of Victoria – Ballaarat Sept 1858

MS 10416 MSB 607, Mundy, Henry, 1831-1912, reminiscences [manuscript]

MS 11484 Box 1777/4, Lazarus, Samuel, diary 24 September 1853 – 21 January 1855

MS 11491 Box 33/5, Reverend Patrick Smyth, letter to W.H. Archer, 13 December 1854

MS 11646 BOX 2178/4-5, Pierson, Thomas, diaries, [manuscript]

MS 11951 Box 2493/3, Ship Diary of James Ashburner. 1876

MS 6167 Box 94/4 (b), Charles Pasley, letters to his father, 1853-1861

MS 7725 Box 646/9, Huyghue, S.D.S, *The Ballarat Riots 1854*

MS 8081 BOX 956/3(d), Stiles, Mr, notes [manuscript]

MS 9370 MSB 567/1-2, Pearce, Harry Hastings, 1897-1984, papers [manuscript]

PROV, VA 466, Governor, VPRS 4066/PO Inward Correspondence, Unit 1, No 69. Ballarat Reform League Charter

VPRS 1085/P Unit 8, Duplicate 162 Enclosure no. 9: Report of Charles P. Hackett to His Excellency the Lieutenant Governor, 3 December 1854

VPRS 1085/P Unit 8, duplicate 162 Enclosure no. 7: Captain Thomas' Report on the Attack on the Eureka Stockade to Major General Robert Nickle, 3 Dec 1854

VPRS 1189 Box 92 54/J14458, Robert Rede to Chief Commissioner, 27 November, 1854

VPRS 1189 Box 92 54/J14460, Robert Rede to Chief Commissioner, 28 November, 1854

VPRS 1189 Box 92 54/J14461, Robert Rede to Chief Commissioner, 1 December, 1854

VPRS 1189 Box 92 J54 14462, Colonial Secretaries Correspondence, Robert Rede's comments regarding the need to crush democratic agitation

VPRS 1189 Box 92 K54/11826, Robert Rede to Chief Commissioner, 22 October, 1854

VPRS 1189 Box 92 K54/13.219, Robert Rede to Chief Commissioner, 25 November, 1854

VPRS 1189 Box 92 K54/13.510, Robert Rede to Chief Commissioner, 30 November, 1854

VPRS 1189 Box 92 K54/13.511, Charles Pasley to the Lieutenant Governor, 29 November, 1854

VPRS 1189 Box 92 K54/13512, Charles Pasley to the Lieutenant Governor, 30 November, 1854

VPRS 1189/P, J54/12058, Captain MacMahon reports on plans for the defence of the Government Camp

VPRS 4066/P Unit 1, December 1854 no. 55, letter from a 'Young Englishman' to Charles Hotham, 4 December 1854

VPRS 4775 Unit 25, Map of Ballaarat Township

VPRS 5527/P Unit 2, Item 2, Private Patrick O'Keefe's trial deposition against John Joseph

VPRS 6927, Eureka Historical Records Collection

Ballarat Archives

Disturbances at Ballarat, 5th December 1854, original document held at the Gold Museum Ballarat

First Report From the Select Committee of the Legislative Council on the Ballarat Outbreak Petition, 12th March 1856, original document held at the Gold Museum Ballarat

Frederick Vern's handwritten plan of the organisation for the Diggers of Ballart, original document held at the Gold Museum Ballart, accession number 78.972

Diary of The Reverend Theophilius Taylor, 23 September, 1853 – 1 August, 1856 (Australia) Ballarat Genealogy Society

The Eureka Stockade – By One of the Insurgents, St Ives Library, Sovereign Hill, Ballarat

The Family of Hyman and Augusta Levinson from Mrs Elias Abraham, St Ives library Sovereign Hill, Ballarat

United Kingdom National Archives

CO309/28, Charles Hotham's Military report to the Colonial Secretary 22 December, 1854

PRO30/22/12C, Letters from Sir George Grey, Colonial Secretary, to Lord John Russell regarding the events at Ballarat

WO12/2971, Muster Rolls, 12th Regiment

WO12/5366, Muster Rolls, 40th Regiment

BOOKS

Adcock, W.E., *The Gold Rushes of the Fifties*, Poppet Head Press, Victoria, 1982

Alcock, Rutherford, *Notes on the Medical History and Statistics on the British Legion of Spain; comprising the results of gun-shot wounds in relation to important questions in surgery*, John Churchill, London, 1838

Anderson, Hugh, *The Colonial Minstrel*, F.W. Cheshire, Melbourne, 1960

____ (ed), *Eureka, Victorian Parliamentary Papers, Votes and Proceedings 1854-1867*, Red Rooster Press, 1999

____ (ed), *The Goldfield Commission Report*, Red Rooster Press, Ascot Vale, Melbourne, 1978

Annear, Robyn, *Nothing but Gold, The Diggers of 1852*, The Text Publishing Company, Melbourne, 1999

Axelrod, Alan, *The Quotable Historian*, McGraw Hill, New York, 2000

Baldwin, J.W., *A Norfolk Soldier in the First Sikh War, Experiences of a Private From H.M. 9th Regiment of Foot in the Battles for the Punjab, India 1845-6*, Leonaur Ltd, London, 2005

Bancroft, Hubert Howe, *The Works of Hubert Howe Bancroft, Vol XXI History of California Vol IV, 1840-1845*, A.L. Bancroft and Company Publishers, San Francisco, 1886

Barnes, R. A., *History of the Regiments and Uniforms of the British Army*, Sphere Books, London, 1972

Bate, Weston, *Lucky City, The First Generation at Ballarat 1851-1901*, Melbourne University Press, Melbourne, 1979

____ *Victorian Gold Rushes*, McPhee Gribble/Penguin Books, Fitzroy, 1988

Bently, N (ed). Russell, W.H., *Russell's Despatches from the Crimea 1854-1856*, Andre Deutsch Ltd, London, 1966

Blackett, R.J.M., *With Divided Hearts – Britain and the American Civil War*, Louisiana State University Press, Baton Rouge, 2001

Blainey, Geoffrey, *The Rush that Never Ended – A History of Australian Mining*, Melbourne University press, Melbourne 1989

Blake, Les, *Peter Lalor, The Man From Eureka*, Neptune Press, Belmont Victoria, 1979

Boessenecker, John, *Gold Dust and Gunsmoke*, John Wiley & Sons Inc, New York, 1999

Bowden, Keith McCrae, *Gold rush Doctors at Ballarat*, Magenta Press Pty Ltd, Melbourne, 1977

Bradfield, Raymond, *Some 'Yankies' on the Central Victorian Goldfields*, Self Published - Castlemaine Education Centre.

Buchanan, Russell A., *David S. Terry of California Duelling Judge*, The Huntington Library, San Marino, California, 1956

Cannon, Michael, *Who's Master? Who's Man? Australia in the Victorian Age: 1*, Thomas Nelson Australia Pty Ltd, West Melbourne, 1978

Carboni, R., *The Eureka Stockade*, The Miegunyah Press, Carlton, 2004

Carrodus, G., *Gold, Gamblers & Sly Grog: Life on the Gold Fields 1851 – 1900*, Oxford University Press, Melbourne, 1981

Castle, J., Close, J., Pryor, G., Willis, R., *Issues in Australian History*, Longman Cheshire, Melbourne, 1982

Clark, F.D., *The First Regiment of New York Volunteers*, Geo. Evans & Co. New York, 1882

Clark, Manning, *A Short History of Australia*, New English Library, London, 1963

Cook, Hugh, *The Sikh Wars 1845-6, 1848-9*, Leo Cooper, London, 1975

Corfield, John, Wickham, Dorothy, Gervasoni, Clare, *The Eureka Encyclopaedia*, Ballarat Heritage Services, Ballarat, 2004

Currey, C.H., *The Irish at Eureka*, Angus and Robertson, Melbourne, 1954

Duncan, J., Walton, J., *Heroes For Victoria*, Spellmount Ltd, Kent, 1991

Dutton, Geoffrey, *S.T. Gill's Australia*, Mead & Beckett Publishing, Sydney, 1981

Edwards, Michael, *Red Year, The Indian Rebellion of 1857*, Cardinal (Sphere Books), London, 1973

Elting, John R., *Amateurs To Arms! A Military History of the War if 1812*, Da Capo Press, New York,

Eureka Reminiscences, Ballarat Heritage Services, Ballarat, 1998

Featherstone, Donald, *Victorian Colonial Warfare – India*, Blandford, London

____ *Weapons and Equipment of the Victorian Soldier*, Arms and Armour Press, London, 1996

Fels, Marie Hansen, *Good Men and True, The Aboriginal Police of the Port Phillip District 1837 – 1853*, Melbourne University Press, 1988

Ferguson, Charles D., *The Experiences of a Forty Niner during a Third of a Century in the Gold Fields*, First Edition 1888, Reprinted by H.A. Garson, Chico California, 1923

____ *Experiences of a Forty Niner in Australia and New Zealand*, Gaston Renard Publisher, Melbourne, 1979

Fitch, Franklin Y., *The Life Travels and Adventures of an American Wanderer, A Truthful Narrative of Events in the Life of Alonso P. DeMilt*, John. W. Lovell Company, New York, 1883

Fitzpatrick, Brian, *The British Empire in Australia, An Economic History 1834 – 1939*, Melbourne University Press, Melbourne, 1949

Fox, Len, *Eureka and its Flag*, Mullaya Publications, Cantebury, Victoria, 1973

Game, C (ed), *We Swear by the Southern Cross: Investigations of Eureka and its legacy to Australia's democracy. State of Victoria*, The Department of the Premier and Cabinet, Carlton South, 2004

Gammage, Bill, *The Broken Years*, Penguin Books, Ringwood Victoria, 1982

Garden, D., *Victoria A History*, Thomas Nelson Australia, Melbourne, 1984

Gilbert, P.F., *Gold*, The Jacaranda Press, Brisbane, 1970

Gold, G., *Eureka – Rebellion Beneath the Southern Cross*, Rigby Limited, Melbourne, 1977

Goodman, David, *Gold Seeking Victoria and California in the 1850s*, Allen and Unwin Pty Ltd, St Leonards NSW, 1994

Grassby, Al, Hill, Marji, *Six Australian Battlefields*, Allen and Unwin, St Leonards, NSW, 1988

Greenway, John, *The Last Frontier – A Study of the Development of the Last Frontiers of America and Australia*, Lothian Publishing Company Pty Ltd, Melbourne, 1972

Grey, Elizabeth, *The Noise of Drums and Trumpets*, Longman Group Ltd, London, 1971

Gurry, Tim (ed), *The European Occupation*, Heinmann Educational Australia, Richmond, 1984

Hale, F., *Wealth Beneath the Soil*, Thomas Nelson Australia, Melbourne, 1981

Hamley, Sir Edward Bruce (General), *The Operations of War, Explained and Illustrated*, William Blackwood and Sons, Edinburgh, 1914

Harvey, Jack, *Eureka Rediscovered*, University of Ballarat, Ballarat, 1994

Historical Studies Australian and New Zealand, Eureka Supplement, Melbourne University Press/Cambridge University Press, London/New York, 1965

Hocking, Geoffrey, *The Red Ribbon Rebellion, The Bendigo Petition 3rd – 27th of August 1853*, New Chum Press, Castlemaine, 2001

_____ *To The Diggings*, Thomas C. Lothian Pty Ltd, Port Melbourne, 2000

_____ (ed), *Early Castlemaine – A Glance at the Stirring Fifties*, New Chum Press, Bendigo, 1997

_____ *Eureka Stockade, the events leading to the attack in the pre dawn of 3 December 1854*, Five Mile Press, Rowville Victoria, 2004

Holmes, Richard, *Firing Line*, Pimlico – Random House, London, 1985

_____ *Redcoat, The British Soldier in the Age of Horse and Musket*, Harper Collins, London, 2002

Howitt, William, *Land, Labour and Gold – Or two Years in Victoria with Visits to Sydney and Van Diemen's Land*, First published in 1855 by Longman, Brown, Green and Longmans, this edition published in 1972 by Lowden Publishing Company, Lowden's Road Kilmore

Hughes, B.P., *Firepower – Weapons Effectiveness on the Battlefield, 1630-1850*, Arms and Armour Press, London, 1974

Hughes, Robert, *The Fatal Shore*, Pan Books, London, 1987

Johnson, Laurel, *Women of Eureka*, Historic Montrose Cottage and Eureka Museum, Ballarat, 2002

Keegan, John, *The Face of Battle – A Study of Agincourt, Waterloo and The Somme*, Barrie and Jenkins, London, 1976

Keir D.L., Lawson F.H., *Cases in Constitutional Law*, Oxford Clarendon Press, London, 1928

Kelly, W., *Life in Victoria or Victoria in 1853 and Victoria in 1858*, reprint of 1859 original, No 6 Historical Reprint Series, Lowden Publishing Company, Kilmore Australia, 1977

Korzelinski, S., *Life on the Goldfields, Memoirs of a Polish Migrant in the ?? 1850's Victoria*, Mentone Educational Centre, Melbourne, 1994

Kruss, Susan, *Calico Ceilings, The Women of Eureka*, Five Island Press Pty Ltd, Melbourne, 2004

Lynch, John, *Story of the Eureka Stockade, Epic Days of the Early Fifties at Ballarat*, (Facsimile) Goldfields Heritage Publications, 1999

MacFarlane, Ian (ed), *Eureka, From the Official Records*, Public Records Office, Arts Victoria, 1995

Mackaness, George (ed), *The Australian Diggings – Where They Are and How to Work Them*, Australian Historical Monographs, Volume VIII (New Series), Review Publications PTY LTD, Dubbo NSW, 1976

McCarthy, Dudley, *Australia in the War of 1939-1945, South West Pacific Area: First Year Kokoda to Wau*, Australian War Memorial, Canberra, 1959

McDougall, Cora (ed), *Gold! Gold!*, Hill of Content Publishing Company Pty Ltd, Melbourne, 1981

McKay, G., *Annals of Bendigo 1851-1920*, Mackay & Co, Bendigo, 1912/1920

McKinnon, S., Stone, D., *Life on the Australian Gold Fields*, Methuen of Australia, Melbourne, 1976

Molony, John, *Eureka*, Viking – Penguin Books Australia, Melbourne, 1984

_____ *History of Australia*, Penguin Books Australia, Melbourne, 1987

Monaghan, Jay, *Australians and the Gold Rush – California and Down Under, 1849 – 1854*, University of California Press, Berkeley, 1966

Myatt, Frederick, *Pistols and Revolvers – An Illustrated History of Hand Guns from the Sixteenth Century to the Present Day*, Tiger Books International, London, 1989

Nosworthy, Brent, *Battle Tactics of Napoleon and his Enemies*, Constable and Company Ltd, London, 1995

_____ *The Bloody Crucible of Courage, Fighting Methods and Combat Experience of the Civil War*, Carroll & Graf Publishers, New York, 2003

O'Brien, Bob, *Massacre at Eureka, The Untold Story*, Sovereign Hill Museums Association, Ballarat, 1973

O'Sullivan, John, *Mounted Police of Victoria and Tasmania*, Rigby, Sydney-Melbourne, 1980

Outbreak at Eureka, The Eureka Story From the Pages of the Mount Alexander Mail, 8 December 1854, Ballarat Heritage Services, Ballarat, 1998

Penzig, Edgar, *Guns and Gold – Stories, Artefacts and Crime of the Australian Diggings 1850-1900*, Tranter Enterprises, Katoomba, 1993

Police in Victoria 1836 – 1980, Victoria Police Management Services Bureau, 1980

Potts, A., Potts, E.D., *A Yankee Merchant in Gold Rush Australia – The Letters of George Francis Train 1853 – 55*, William Heinemann Australia, Melbourne, 1970

_____ *Young America and Australian Gold – Americans and the Gold Rush of the 1850s*, University of Queensland Press, St Lucia, 1974

Price. B.J., *The Australian Gold Rushes*, Reed Education, Sydney, 1972

Quaife, G., *Gold and Colonial Society 1851-1870*, Carsell Australia Ltd, Melbourne, 1975

Records of the Castlemaine Pioneers, Rigby Limited, Melbourne, 1972

Ritchie, Robert Welles, *The Hell Roarin' Forty Niners*, J.H. Sears and Company Inc, New York, 1928

Sadleir, John, *Recollections of a Victorian Police Officer*, First published by George Robertson & Company 1913, published in facsimile edition by Penguin Books, Blackburn Victoria, 1973

Serle, Geoffrey, *The Golden Age – The History of the Colony of Victoria 1851-1861*, Melbourne University Press, Parkville, 1963

Shaw, A.G.L., Nicholson H.D., *An Introduction to Australian History*, Angus and Robertson Ltd, Sydney, 1962

Smith, J. Graham, *Reminiscences of the Ballarat Gold Field*, Pick Point Publishing, Ballarat, 2002

Smith, Neil. C., *Soldiers Bleed Too – The Redcoats at the Eureka Stockade*, Mostly Unsung Military History, Melbourne, 2004

Smyth, R.B., *The Gold Fields and Mineral Districts of Victoria*, Queensbury Hill Press, Carlton, 1980 (reprint of the original 1869 edition)

Smythies, R.H., *Historical Records of the 40th (2nd Somersetshire) Regiment now 1st Battalion The Prince of Wales's Volunteers (South Lancashire Regiment) From its Formation, 1717, to 1893*, A.H. Swiss, Devonport, 1894

Spielvogel, Nathan F.A., *Thrilling Story!!! The Affair at Eureka*, John Fraser & Son, Ballarat, 1928

Stanley, Peter, *The Remote Garrison, The British Army in Australia*, Kangaroo Press, NSW, 1986

Stone, Derrick. L., MacKinnon, Sue, *Life on the Australian Goldfields*, Methuen of Australia, Sydney, 1976

Stoney, Capt. H. Butler, *Victoria: With a description of its principal cities, Melbourne and Geelong: and remarks on the present state of the colony; including an account of the Ballaarat disturbances, and the death of Captain Wise, 40th Regiment*, Smith, Elder & Co. Dublin, 1856

Strachan, Hew, *From Waterloo to Balaclava, Tactics, Technology and the British Army 1815 – 1854*, Cambridge University Press, Cambridge, 1985

Strange, Bert, Strange, Bon, *Eureka, Gold Graft and Grievances*, Bert and Bon Strange, Ballarat, 1973

Tsouras, Peter G., *Military Quotations from the Civil War*, Sterling Publishing Company Inc. New York, 1998

Tuchman, Barbara, *Practising History*, Papermac, A Division of MacMillan Publishers Limited, London, 1991

Turnbull, Clive, *Eureka – The Story of Peter Lalor*, The Hawthorn Press, Melbourne, 1946

Turnbull, Patrick, *Solferino – The Birth of a Nation*, Robert Hale Limited, London, 1985

Turner, G.T., *A History of the Colony of Victoria From its Discovery to its Absorption into the Commonwealth of Australia Vol II. A.D. 1854 – 1900*, facsimile of the 1904 original, Heritage Publications, Melbourne, 1973

Turner, Henry Gyles, *Our Own Little Rebellion – The Story of the Eureka Stockade*, Whitcombe & Tombs Limited, Melbourne, 1913

Urban, Mark, *Rifles – Six Years with Wellington's Legendary Sharpshooters*, Faber and Faber, London, 2003

Walshe, Robert Daniel, *Great Australian Gold Rush & The Eureka Stockade*, Literary Productions Pty Ltd., Janalli, nd

Walshe, R.D., *The Eureka Stockade 1854 - 1954*, Current Book Distributors, Sydney, 1954

Weapons, An International Encyclopaedia from 5000 B.C. to 2000 A.D, The Diagram Group, Macmillan, London, 1980

Wickham, Dorothy, *St Alipius Ballarat's First Church, The Early History*, D. Wickham, Ballarat, 1997

_____ *Deaths at Eureka*, D, Wickham, Ballarat, 1996

Wickham, Dorothy; Gervasoni, Clare, D'Angri, Val, *The Eureka Flag Our Starry Banner*, Ballarat Heritage Services, Ballarat, 2000

Wickham, Dorothy; Gervasoni, Clare, Phillipson, Wayne, *Eureka Research Directory*, Ballarat Heritage Services, 1999

Williams, Vic (ed), *Eureka and Beyond – Monty Miller, His Own Story*, Lone Hand Press, Perth, 1988

Winders, R.B., *Mr. Polk's Army – The American Military Experience in the Mexican War*, Texas A&M University Press, College Station, 1997

Withers, William Brawell, *History of Ballarat*, Facsimile Edition of 1887 original, Queensbury Hill Press, Carlton, Victoria, 1980

JOURNALS-MAGAZINE ARTICLES

'*Grandma hid miner under her skirts at Eureka*', The Ballarat News, 15 June 1983

Ballarat Link, The Ballarat & District Genealogical Society Inc., November 2004 – No.143

Birell, Ralph, *Spying on the Goldfields*, Victorian Historical Journal, Vol. 66, No 1, June 1995

Blainey, Geoffrey, *Victoria's Bloody Sunday*, Royal Auto Magazine, November 2004

Bowie, Patrick, *The Shifting Gold Rush Scenario: California to Australia and New Zealand*, The Californians, The Magazine of California History, Vol. 6 number 1, Jan/Feb 1988

Fyson, Anthony, *Eyewitness at Eureka*, History Today, Vol. 54 (12) December 2004

Ireland, John, *Eureka: Politics or Self Defence*, Victorian Historical Journal, Vol. 68, No 1, April 1997

Kirkby, Dianne, *Gold and the Growth of Two Metropolis: a Comparative Study of San Francisco and Melbourne, Australia*, Journal of the West, April 1978

Linane, Father T.J., *Names in the Eureka Story*, 'Light', September 1977

Magee, R., *Muskets, Musket Balls and the Wounds They Made*, Australian New Zealand Journal of Surgery, (1995) 65

Manne, Robert, *A Turkish Tale: Gallipoli and the Armenian Genocide*, The Monthly, February 2007

McKernan, Christine, *40th Regiment at Eureka, 12th Regiment, 1st Battalion at Eureka*, Ancestor Magazine, Vol. 23 No. 7 Spring, 1997 – Genealogical Society of Victoria

McMullin, R., *The Impact of Gold on Lawlessness and Crime in Victoria 1851-1854*, The Victorian Historical Journal, Vol. 48, May 1977

Michael Carroll's Memoirs – 2, Investigator – Geelong Historical Society, June 1999

Molony, John, *Eureka and The Australian Republic, Foreword to Anderson, H (ed) Eureka, Victorian Parliamentary Papers, Votes and Proceedings 1854-1867*, Red Rooster Press, 1999

_____ *Eureka and the Prerogative of the People*, a paper presented in the Senate Occasional Lecture series 23 April 2004.

Nafziger, G., *Skirmisher tactics of the Napoleonic Wars, Part 1: British Skirmishers*, Empires, Eagles, & Lions, 14

U.S. Army Field Artillery School Fort Sill, Oklahoma. *Battle Analysis Instruction Booklet*

Warren, V.L., *'Dr John Griffin's Mail 1846-1853'*, California Historical Society Quarterly, volume XXXIII, number 3, September 1954

UNPUBLISHED MANUSCRIPTS

Sunter, Anne Beggs, *Birth of a Nation? Constructing and Deconstructing the Eureka Legend*, Doctor of Philosophy Submission, April 2000, Department of History, The University of Melbourne

The Siting of the Eureka Stockade, Gold Museum Ballarat, August 2004

Walshe, R.D., *The Eureka Tradition – Selections, 1854-1954*, Gold Museum Ballarat, nd

VISUAL REFERENCES

Fostyn, Brian, *The Thin Red Line*, Pimpernel Studios

WEBSITES

Australian Dictionary of Biography – Online: http://www.adb.online.anu.edu.au/adbonline.htm

Ballarat Genealogy Society – Reverend Theophilius Taylor http://www.ballaratgenealogy.org.au/art/t_taylor/1854c.htm

Canadian Military Heritage http://cmhg.gc.ca/cmh/en/page_425.asp

Census of Great Britain 1851, vol. i (1852), Report, Section 8 (Some of the General Results of the Inquiry), lxxxii-lxxxiv: http://ds.dial.pipex.com/town/terrace/adw03/peel/p-health/1851cens.htm

Crimean War Living History 19th Foot (The Green Howards) www.19thfoot.co.uk

Department of History, The University of Memphis, History is … , http://history.

memphis.edu/history_is.html

Eureka on Trial – Map http://eureka.imagineering.net.au/www.site/access_version/map. htm?fl=t

Fort Henry Home Page: http://collections.ic.gc.ca/fort_henry/FortHome.htm

George Mason University-History News Network http://hnn.us/articles/1328.html

Gold – Victorian Cultural Collaboration http://www.sbs.com.au/gold/story. html?storyid=79#1555

Greek Tragedy: http://cs.clark.edu/~hum101/Humanities_101/greek_tragedy.htm

Land Forces of the British Empire: http://www.britishempire.co.uk/forces/ armyarmaments/rifles/sniderhistory.htm

Mark Twain Quotes. http://www.twinquotes.com/Guns.html

Mark Twain: A Biography, Chapter 11 Days of Education. http://extext.library.Adelaide. edu.au/t/twain/mark/paine/chapter11.html

Mess Tent: http://users.hunterlink.net.au/~ddchr/MESS%20TENT.htm

MI6 AR Rifle: http://www.hk94.com/m16-rifle.html

Monty Miller: http://www.cpa.org.au/garchve04/1209eureka.html

New York Volunteers http://www.mtdemocrat.com/articles/2005/01/24/columnists/ doug_noble/y1401_d.txt

Police Force History: http://members.westnet.com.au/talltrees/vicpol/victoria.htm

Reclaiming The 'Freedom' Heritage of Eureka stockade. http://ausfirst.alphalink.com.au/ articles/partthree.html

Regimental School System in the British Army in the Napoleonic Era:http://www. napoleonseries.org/military/organization/c_rgtschool1.html

Sniper's Paradise: http://www.snipersparadise.com/history/sharps.htm

The Avalon Project at Yale Law School, English Bill of Rights: http://www.yale.edu/ lawweb/avalon/england.htm

The Avalon Project at Yale School of Law: An Act Declaring the Rights and Liberties of the Subject and setting the Succession of the Crown.http://www.yale.edu/lawweb/avalon/ england.htm

The Guardian December 1, 2004 Reminiscences of the Eureka Stockade: http://www.cpa. org.au/garchve04/1209eureka.html

The System of Purchasing One's Rank and Position: http://www.cwreenactors. com/~crimean/purchsys.htm

The United States Navy Observatory http://aa.usno.navy.mil

NEWSPAPERS

Argus

Ballarat Courier

Daily Alta California

Mount Alexander Mail

ENDNOTES

Introduction

1. MS 10416 MSB607 Vol 3, p546.
2. Argus 11 December 1854
3. Hotham, in his report to the Colonial Secretary, refers to a soldier being killed outright by the first shots from the stockade. According to the military casualty returns for Eureka Private Michael Roney of the 40th Regiment was killed by a shot through the head. It is quite likely he was the man Hotham referred to as at no other time did the official accounts report, inaccurately as it would transpire, any other soldier being killed outright during the battle.
4. Eureka Reminiscences, p44.
5. MS6167 Box 9414(b) 55-57. Captain Charles Pasley's reference to a 'trifling affair' needs to be understood in the context of a professional soldier's expectations especially at a time when war was raging in the Crimea.
6. Keegan, p103.

Chapter 1

1. Ferguson, p 55.
2. In August 1853 tens of thousands of miners on the Bendigo diggings organised mass action against the manner in which the gold fields were being administered. This would become known as the 'Red Ribbon Rebellion' from the red ribbons the protesting miners wore. A formal petition signed by thousands of Bendigo miners was submitted to the colonial government of the day, and received no meaningful response. Exasperated, many began to arm themselves, some even going to the extent of stripping the lead from the wooden chests merchants kept in their stores to use to make ammunition. A mass protest succeeded in attracting almost ten thousand miners, many of whom carried firearms and formed up under their respective national flags. Eventually the situation was defused by the tact and common sense of the Resident Commissioner, Joseph Panton.
3. *Records of the Castlemaine Pioneers*, p 68.
4. *Argus* 8 December 1854.
5. PROV, VA 466 Governor, VPRS 4066/PO Inward Correspondence, Unit 1, No 69.
6. Governor Hotham paid little attention to the Charter when it was presented to him by a delegation from the Reform League preferring instead simply to write 'Put away' on it.
7. Egan was wounded but eventually recovered. His wounding sparked a persistent myth that he had died, which remained in vogue for many decades until disproved by Ballarat Historian Dorothy Wickham.
8. Historical Studies, Australia and New Zealand - Eureka Supplement, p 25.
9. Serle, pp183-186.
10. The Australian League of Rights: http://www.alor.org/Volume9/Vol9No46.htm.
11. See Karl Marx's commentary on Eureka in Corfield, Wickham and Gervasoni, p 366 – 367.
12. *Anarchist Age Weekly Review*: http://www.ainfos.ca/02/dec/ainfos00059.html.
13. Molony in Anderson p viii - xii, Grassby, Hill, pp200-238.
14. http://www.hinch.net/says_archive04/Nov04/19-11-04.htm.
15. *Argus* 19 December 1854.

16. Australian Nationalism Information Database - www.ozemail.com.au/~natinfo.

17. Corfield, Wickham and Gervasoni, p95.

18. Molony p 108.

19. Corfield, Wickham and Gervasoni, p186.

20. Bate, p 46.

21. Molony, p108.

22. Serle, p181,

23. Hughes, pp564-565.

24. Carrodus, p16.

25. John Adams: Essay on the Canon and Feudal Law. http//www.colonialhall.com/adamsj/adamsj2.asp

26. *Argus* 2 October 1854

27. Historical Studies, Eureka Supplement p 28-29.

28. http://www.yale.edu/lawweb/avalon/england.htm.

29. Lalor in the *Argus*, 10 April 1855.

30. Eureka Reminiscences, p21.

31. Eureka Reminiscences, p61.

32. MS 11646 Box 2178/4-5.

33. Carboni, p17.

34. Parliamentary Papers, p158.

35. Ferguson, p 52.

36. *Argus*, 2 October 1854.

37. *Argus*, 8 Dec 1854.

38. *Argus*, 6 December 1854.

39. *Argus*, 10 April 1855.

40. Quaife, p179.

41. Potts and Potts p179.

42. MS 11646 Box 2178/4-5 p 237.

43. MS 9370 MSB576/1-2 Harry Hastings Pearce Papers, p20.

44. Keir and Lawson, p371.

45. Keir and Lawson, p371.

46. Blake, p54. Whether or not the military did in fact fire being irrelevant to the passions being expressed at the time.

47. *Argus*, 8 December 1854.

48. *Argus*, 2 December 1854.

49. Withers, p96.

50. Withers, p79.

51. Some notable examples of the army being used to suppress civil disorder during these times are the Gordon riots of 1780, the subjugation of the Luddites in 1812, the Peterloo Massacre of 1819, the Bristol riots of 1831 and the bloody repression of Chartism in Newport and other towns during the 1830s. There was also the frequent use of the army to maintain order in Ireland.

52. Carboni, p6.

53. Carboni, p65.

54. Quaife, p78.
55. Carboni, pp129-131.
56. Anderson, pvii.
57. Withers, p66.
58. Carboni, p5.
59. Corfield, Wickham and Gervasoni, p302.
60. Queen v. Joseph, p48.
61. Carboni, p59.
62. Wickham, Gervasoni, D'Angri, p17.

Chapter 2

1. Queen v. Joseph p14.
2. Ferguson, p53. The use of improvised pikes by civilian rebels was common during the eighteenth and nineteenth centuries. During the great rebellion of 1798, Irish rebels adopted the pike and used it in numerous battles with British troops, as well as in civil disturbances throughout the land. Chartist rebels in England and nationalist rebels in Wales used the pike during the 1830s. In nearly all cases the pike proved to be ineffective against troops armed with firearms. Yet in circumstances where access to firearms was limited, the pike proved to be cheap, quick and easy to produce, required minimal maintenance, and was simple to use, hence its allure for hastily raised armies of militia such as that of the insurgent miners at Eureka.
3. *Eureka Encyclopaedia*, p390.
4. Fitzpatrick, p207, Strange, p19, Molony, p149.
5. Carrodus, p20.
6. *Argus,* 1 Dec 1854.
7. Carboni, p55.
8. Clancy in Turner, p119.
9. Outbreak at Ballarat, p12.
10. MS11484 Box 1777/4.
11. MS7725 Box 646/9.
12. Issues in Aust History, p22.
13. Lalor to the *Argus*, 10 April 1855 in Historical Studies, Eureka Supplement, p38.
14. Carboni, p79.
15. Lynch, pp29-30.
16. Lalor to the *Argus*, 10 April 1855.
17. *Eureka Reminiscences* p33.
18. *Argus*, 9 Feb 1852, Potts and Potts, p166.
19. Greenway, p23.
20. Corfield, Wickham and Gervasoni, p239.
21. Featherstone, *Weapons and Equipment of the British Soldier*, pp49-50.
22. Ferguson, p62.
23. See Boessenecker *Gold Dust and Gunsmoke* for an excellent account of life on the Californian gold diggings during the late 1840s and into the 1850s.
24. Boessenecker, p321.

25. Goodman, p69.

26. Carboni, p85.

27. Kelly, vol.1, p180.

28. Potts and Potts, p170

29. Carrodus, p 78.

30. Pierson, pp139-141 and pp239-249, MS 11646 Box 2178/4-5.

31. Johnson, p35.

32. Records of the Castlemaine Pioneers, p157.

33. http://www.ballaratgenealogy.org.au/art/t_taylor/1854c.htm.

34. Lynch, p31.

35. Withers, p109.

36. Lalor's pistol is held in the collection of the State Library of Victoria.

37. *Argus* 4 Dec 1854.

38. VPRS 5527/P Unit 2, Item 9

39. VPRS 5527/P Unit 2, Items 2 and Queen v. Joseph.

40. Lynch, p31. VPRS 5527/P Unit 2, Item 5.

41. Queen v. Hayes, p76. *Argus*, 11 December 1854.

42. Queen v. Hayes, p79. Queen v. Joseph, p13.

43. Stone and MacKinnon, p76.

44. Carboni, p68.

45. MacFarlane, p94.

46. Queen v. Hayes, p80.

47. Charles. D. Ferguson's account of the battle for the Eureka Stockade, discussed in more detail elsewhere, mentions Burnette as the man who, using his rifle, fired at and hit a soldier 150 yards from the stockade in the first shot of what Ferguson called 'The Ballarat War'. See Ferguson, p60.

48. VPRS 5527/P Unit 2, Item 2.

49. VPRS 5527/P Unit 2, Item 3.

50. VPRS 5527/P Unit 2, Item 5.

51. Molony, p163.

52. MS11646 BOX 2178/4-5, diaries of Thomas Pierson, pp239-249.

53. Withers, p103.

54. ibid.

55. ibid.

56. ibid.

57. Queen v. Joseph, p27.

58. Carboni, pp62-63.

59. Carboni, p93.

60. Corfield, Wickham and Gervasoni, p12.

61. *Argus* 12 December 1854.

62. MS7723 Box 646/9.

63. Allan, p14.

64. Carboni, p89.

65. In an article written in the *Argus* in July 1909, Dr. W.H. Fitchett mentions the resentment felt by storekeepers because of the demands made on them by the insurgents.

66. Queen v. Joseph, p22.

67. Queen v. Hayes, p77, Queen v. Joseph, p32

68. Withers, 119-120.

69. Carboni, p87.

70. ibid.

71. Miller in the *Guardian* 1 December 2004: http://www.cpa.org.au/garchve04/1209eureka.html.

72. Queen v. Joseph, p27.

73. Queen v. Joseph, p33.

74. Keegan, p154.

75. Keegan, p153.

76. Korzelinski, p45.

77. Lynch, p28-29.

78. Carboni, p58.

79. Carboni, p81.

80. Carboni, p86.

81. Allan, p16. Allan's pedantic description of the company consisting of 'Californians and Americans' is consistent with the language of the day in which a distinction was made between those who had come from the Californian gold fields but who were not native to the United States, and those who were.

82. The original hand drafted letter by Vern is in the collection of the Gold Museum at Ballarat. Written in a clear flowing hand, it reads: '*Plan of Organisation for the Diggers of Ballarat. Let every 7 men select a trust or sub officer, who will be responsible for the immediate appearance if wanted of the 6 men under his comand (sic), 7 of those little detachements (sic) to form a company, and to select a captain. The captain to be responsible for his company, and to keep the muster rolls. 8./ companies to form a brigade. Every company to select 3 members for the electione (sic) of a military commissione (sic) and a commander in chief of the Ballarat forces. The captaine (sic) of every company to appoint a meeting place for the company, every trust officer to appoint a place of meeting for his six men. The commander in chief to appoint a meeting place for the united forces, and to errect (sic) a flagstaff for giving signal to the assemble. F. Vern late aide de camp to Gen Miller.*' Scribbled in the top left hand corner was the following: '*every company to pay 10sh a week to their military commission*'.

83. Corfield, Wickham, Gervasoni, p261.

84. Carboni, p60.

85. Lynch, p13.

86. Carboni, p61-62.

87. Lynch, p15.

88. Carboni, p66.

89. Lynch, pp11-12.

90. Stoney, p126.

91. Carboni, p77, 'Construct barricades!' Carboni misspells the word Barrikaden.

92. Lynch, p12.

93. Carboni, p80.

94. Allan, p15.

95. Withers, p109.

96. Ferguson, pp58, 60.

97. Carboni, p96.

98. *Argus* 10 April 1855.

99. Carboni, pp77, 78, 89.

100. Carboni, p63.

101. Lynch, p13.

102. Withers, p124.

103. Also named Henry Ross in an inscription on a photograph held at the Ballarat Fine Arts Museum.

104. Ferguson, p53.

105. Potts and Potts, *A Yankee Merchant*, p166. Neither Potts and Potts nor Corfield, Wickham and Gervasoni have been able to find any reference to James McGill having attended West Point.

106. Potts and Potts, p186.

107. California Historical Quarterly v.XXXIII, no. 3, September 1954, p 257.

108. Lynch, p29.

109. Carboni, pp57-58.

110. Lynch, p57.

111. Allan, p15.

112. Lynch, p29.

113. *Argus* 10 April 55, Carboni, p88.

114. Shanahan in Withers, p117. Unfortunately Withers does not specify if it was Edward Shanahan or his brother Timothy Shanahan who offered this observation.

115. *Argus*, 10 April 1855.

116. *Argus* 6 December 1854.

117. Lynch, p30.

118. Queen v. Joseph, pp28, 30.

119. Carboni, p90.

120. Carboni, p97.

121. Corfield, Wickham and Gervasoni, p25.

122. Geoffrey Blainey interview on ABC *Lateline*, 7 May 2001.

123. There are numerous accounts of how many insurgents had gathered at Eureka, ranging from 700 to 1500. These men were organised into companies of about 100 men.

124. Carboni, p81.

Chapter 3

125. 'Foot' being the designation used for infantry regiments within the British army at the time.

126. The 40th, in company with two Bengali regiments, attacked a line of 23 Mahratta artillery guns covering the northern side of Maharajapore village. Advancing into withering fire, the colonel of the 40th was killed, as were the next two officers who took his place. On reaching the guns, the men of the 40th went to work with the bayonet, routing the Mahratta gunners who defended their pieces with great courage. In all the 40th lost 140 men including seven officers at Maharajapore. Captain John Thomas, who commanded the attack on the Eureka Stockade, served both in Afghanistan and

at Maharajapore, where he was wounded in the thigh.

127. *Annals of Bendigo*: V1: 28.C.

128. Holmes, p148. Wellington also referred to his soldiers as 'fine fellows' and expressed his admiration for their fighting qualities.

129. MS 7725 Box 646/9.

130. Duncan and Walton, p9.

131. The minimum expenditure to cover all the costs expected of a gentleman for an ensign, the most junior rank within an infantry regiment, was £100 per annum. An Ensign's pay was £96 annum. More senior ranks would be expected to spend substantially more.

132. Duncan and Walton, p9.

133. Duncan and Walton, p45.

134. http://www.cwreenactors.com/~crimean/purchsys.htm.

135. Duncan and Walton, p43. Crimean War hero Sir Colin Campbell was the son of a carpenter, and Field Marshal Sir William Robertson was the son of a country tailor. Three percent of officers were commissioned from the ranks during the Victorian era, and many of these enjoyed illustrious military careers.

136. For an excellent discussion of the debates and changes occurring within the British army during the era 1815-1854, see Scherer, *From Waterloo to Balaclava*.

137. Soldiers such as Privates John Hall, from Kilkenny, Michael Roney, Felix Boyle, Patrick Lynott, John Neil, Daniel Hegarty and John Donnelly are all examples of Irishmen in the ranks.

138. Holmes, p56.

139. Census of Great Britain 1851 http://ds.dial.pipex.com/town/terrace/adw03/peel/p-health/1851cens.htm.

140. The two schools were The Royal Hibernian Military School in Dublin, and The Royal Military Asylum (The Duke of York's School) in Chelsea.

141. Daniel Defoe: *Giving Alms No Charity: Addressed to the Parliament of England*: at http://www.underthesun.cc/Classics/Defoe/almsnocharity/

142. Holmes, p149.

143. Attributed to Lord Wolseley when speaking of the common soldier, Holmes, p25.

144. So called during the reign of Victoria. During reigns of a male monarch it was known as the 'King's Shilling'.

145. Featherstone, D: *The Victorian Soldier*, p1.

146. Hale, pp21-22.

147. Holmes, p300.

148. Holmes, pp294-7.

149. For an excellent description of the role of Chaplains in the British army during the era see Holmes, pp115-120.

150. MS. 11491: box 33/5.

151. The Regimental School System and Education in the British Army in the Napoleonic era – found at http://www.napoleon-series.org/military/organization/c_rgtschool1.html.

152. http://www.bl.uk/collections/early/victorian/pr_intro.html.

153. Holmes, pp115-120.

154. Fort Henry Homepage http://collections.ic.gc.ca/fort_henry/FortHome.htm.

155. In December 1845 during the First Sikh War, the soldiers of the 9th Foot were issued with a drink of brandy on the morning of the second day of the battle of Ferozeshah. This was following a night when the regiment had bivouacked among the carnage in front of the still to be secured Sikh camp, and had been subjected to the deadly attention of enemy artillery throughout the night. J.W. Baldwin, a private soldier with the regiment, described the issuing of brandy and the effect it had of keeping the men's spirits up. Baldwin, p37.

156. *Ballarat Times* Vo2 December 1854.

157. Holmes in *Redcoat* on p325 gives an example of how in 1835 the Mayor of Gloucester refused to allow the 14th Light Dragoons to flog a miscreant soldier within the town limits because of the town's opposition to capital punishment. The Dragoons had to take their man four miles outside the town to do the deed.

158. Holmes, *Redcoat*, p324.

159. Pensions were notoriously small, often being only one shilling a day. If a man was fortunate he could secure a position as a Yeoman of the Guard, the famous 'Beefeaters', or a berth at the Royal Hospital, Chelsea. Otherwise his life could be very grim. Many an old soldier succumbed to drink, became a beggar and lived as best as he could on the streets. The government did virtually nothing to care for its old soldiers. Not until 1855 was an official employment society for former soldiers and sailors set up, and in 1859 the Corps of Commissionaires was established. See: Duncan and Walton, *Heroes For Victoria*.

160. O'Brien, p120.

161. AJCP M2579.

162. Duncan, Walton, p43.

163. Duncan, Walton, p11.

164. O'Brien, p67.

165. *Eureka Encyclopaedia* p527.

166. *Eureka Reminiscences*, p58.

167. Lynch, pp32-33. Carboni also mentions Harris being kind the prisoners, but gives it a bitter twist when he describes Harris's as being 'officious' and condescending in his 'affection'. It was Carboni's impression that Harris was intent on humiliating some of the prisoners. One suspects that Carboni's own humiliation at the manner in which he had been arrested and chained with the other prisoners is the root cause of his animus.

168. Smith, pp141-142.

169. Queen v. Joseph, p 20. The jacket Lynott was referring to was the 1846 pattern undress shell jacket, which according to the 1846 dress regulations modified in 1848 was a scarlet, singled breasted jacket with nine or ten pewter regimental buttons, plain collar, cuffs and shoulder straps of facing colour. It was far more practical and comfortable than the heavy, long tailed woollen coatee.

170. MS 7725 Box 646/9.

171. *Eureka Stockade* by Samuel Douglas Smyth Huyghue, Ballarat Fine Art Gallery.

172. Desmond O'Grady in *Raffaello*, pp.157-158, gives a good account of the conditions experienced by the troops marching up to Ballarat from Melbourne.

173. The deleterious effects of wearing the British Army knapsack issued during this era are described in detail in Featherstone, D., *Weapons and Equipment of the Victorian Soldier* p114.

174. Lynch, J, pp 32-33.

175. So called after George Lovell who was appointed Inspector General of Small Arms in 1840.

176. See Appendix 1

177. Holmes, R., *Redcoat*, p394.

178. Holmes, p76.

179. Carboni, p98.

180. O'Sullivan, p78.

181. The original painting is to be seen at the Ballarat Fine Art Gallery.

182. *Joes*, so called after Governor Joseph Latrobe.

183. *Argus* 2 October 1854.

184. In a very short time, 40 out of the 50 City Police in Melbourne had resigned. Fifteen Country Police did the same. In an attempt to stem the tide, the colonial government raised police pay to 6 shillings a day, but this had little effect and another five constables resigned soon after.

185. In 1852 the colony was divided into Police Districts for the first time, but this did little to coordinate the City Police, Goldfields Police, Water Police, Rural Bench Constabulary, Mounted Police, Country Police, Escort Police and Aboriginal Police, all of whom were operating within Victoria at the time.

186. One such miscreant was the notorious Superintendent David Armstrong, the so called Flying Demon, who as well as freely indulging in bribery, delighted in cracking miners' heads with his metal knobbed whip handle. Police Magistrate John D'Ewes and police Sergeant Major Milne were both dismissed following a public outcry and official investigation for corrupt practices following the Bentley trial. See Corfield, Wickham, Gervason p 151, p 377.

187. O'Sullivan, p57.

188. Such weapons were certainly used by some of the police. O'Sullivan, pp58-59, mentions a parade of the mounted police in Melbourne on 2 December 1853. One police unit was armed with 'new improved American carbines'. This carbine was described as being a weapon that 'cut off the cartridge ends automatically and supplied itself with fifty caps thus avoiding the trouble of putting caps on nipples'. Being an American carbine, and described in such a way, this could only have been the Sharpes carbine fitted with the Maynard Cap roll system which allowed the continual feeding of ignition caps to the firearm's nipple rather than having to place a cap. The Sharpes carbine, introduced in 1848, was a good robust weapon, which, as its barrel was rifled, had a longer effective range than the smooth bore carbines normally carried by the police.

189. In his testimony at the State Trials in 1855, Queen v. Joseph, Amos made the comment that even though a trooper may carry a pistol holster, it did not follow that there was a pistol inside it. In a show of pique, verging on insolence, he replied to a question from the defence counsel regarding the carriage of pistols in the holsters by the police by saying that, as he could not see through the leather of the holsters, he could not swear that the troopers were carrying pistols inside them. He then added that many troopers carried pocket pistols.

190. Sketch held at the Victoria Police Historical Unit, Police Headquarters, Flinders Street Melbourne.

191. http://www.art-museum.unimelb.edu.au/victorian_gold/life/gill09.html.

192. King's shotgun can now be seen displayed at the Eureka Centre Museum in Ballarat.

193. Corfield, Wickham and Gervasoni, p307.

194. MS 7725 Box 646/9

Chapter 4

1. Book titles such as *Massacre at Eureka – the Untold Story* characterise such approaches to telling the Eureka story.

2. Carboni, p70.

3. MS11646 Box 2178/4-5 p239-249.

4. Allan, p14.

5. Lynch, p30.

6. Queen v. Joseph pp13, 15.

7. Carboni, p81.

8. Withers, p105.

9. VPRS 6927.

10. Pierson, pp239-249, MS 11646, Box 2178/4-5.

11. Carboni, pp79-80.

12. Ferguson, p58.

13. MS 11646, Box 2178/4-5 p 247.

14. Carboni, p89.

15. Carboni, p85.

16. MS6167 Box 9414 (b).

17. Mackay, p28.

18. *Records of the Castlemaine Pioneers*, p6.

19. *Records of the Castlemaine Pioneers*, pp100-101.

20. Howitt, pp321-322.

21. *Records of the Castlemaine Pioneers*, pp100-101.

22. Carboni, pp80-81.

23. Allan, p15.

24. Queen v. Joseph, p36, Queen v. Hayes, p86.

25. *Eyewitness at Eureka*, History Today, December 2004, p33..

26. Carboni, pp80-81.

27. Queen v. Joseph, p23.

28. Queen v. Joseph, p27.

29. Queen v. Joseph, p29.

30. Queen v. Joseph, p29.

31. Ibid.

32. Queen v Hayes, p77.

33. Queen v. Hayes, p77.

34. Queen v. Joseph, p21.

35. Queen v. Joseph, p20.

36. Harvey, p88.

37. The *cheveaux de frise* is an ancient technique where sharpened stakes are placed in the ground angled outwards. This is done to counter mounted enemies and to break up the formations of attacking infantry. The builders of the Eureka Stockade may or may not have considered this when they constructed their defences, although Huyghue's description of some of the slabs being cut to points, if it was factual, could indicate that they did.

38. Crossing the stockade's slab palisade was not an easy task for the attackers. Many soldiers, unable to push it down, jumped on top, and some fell. See Queen v. Joseph, p20.

39. Monaghan, p258.

Chapter 5

1. MS 6167 Box 9414 (b).
2. VPRS 6927.
3. Huyghue in O'Brien, p14.
4. Cusack, p105.
5. VPRS 1189 Box 92 54/J14462.
6. VPRS 1085/P Unit 8, duplicate 162 Enclosure no.7
7. VPRS 1189/P Unit 92, J54/12058, p11.
8. MS 6167 Box 9414 (b).
9. Withers, p122.
10. Withers, p123.
11. Withers, p104.
12. Carboni, p72.
13. Carboni, p77.
14. MS 6167 Box 9414 (b), p50.
15. VPRS 1189 BOX92 54/J14460.
16. VPRS 1189 Box 92 K54/13. 511.
17. MS 6167 Box 9414 (b), p50.
18. VPRS 1189 Box 92 K54/13512.
19. VPRS 1189 Box 92 K54/11826.
20. VPRS 1189 Box 92 J54/11826, J54/11.896, J54/12.107.
21. VPRS Box 92 55/J14458.
22. VPRS 1189/P, J54/14459.
23. VPRS 1189 Box 92 K54/13.219, 54/J14460.
24. VPRS 1189 Box 92 K54/13. 510.
25. VPRS 1189 Box 92 54/J14462.
26. MS 6167 Box 9414 (b).
27. Historical Studies – Eureka Supplement, p 29.
28. AJCP M2579: the records of the 40th Regiment for 1855 list 400 men joining the regiment from England and 100 joining as volunteers from the 99th Regiment, then in Tasmania. During 1855 the regiment lost 76 men for various reasons, giving it 980 effectives on 31 December 1855. This was a marked increase from the 582 effectives recorded for December 1854. PRO30/22/12C f.49 the letters of Sir George Grey, Colonial Secretary, to Lord John Russell regarding the events at Eureka indicate that 150-200 men of the 60th Regiment had been sent to Australia in March 1855. Even if not stationed in Victoria, these troops would be close to hand if needed there. CO309/28, a military report of 22 December 1854 from Hotham to Grey, requests that the 12th Regiment be retained in Victoria, and a battery of 32 pounder rockets be sent to the colony.
29. Cunningham, p78-79.
30. Cunningham, p79.
31. Hocking, *To the Diggings*, p22.
32. Axelrod, p43.
33. Blackett, p14.

34. Blackett, pp13-14.

35. Goodman, p71.

36. Bob O'Brien in *Massacre at Eureka* makes a great deal of the clandestine communications between Hotham and Rede, stating forthrightly that the enciphered messages were direct orders from Hotham to crush the insurgents with military force and massacre them. As the contents of those messages remain unknown and existing correspondence from Hotham made the point that temper and moderation were to be used when dealing with the dissident miners, such claims as those made by O'Brien are unlikely.

37. See Appendix 2.

38. VPRS 3219/P Unit 2, 3430.

39. See O'Brien: Massacre at Eureka.

40. Maharajapore, in particular, was a very hard fought battle in which the 40th lost its colonel, as well as 174 men killed and wounded when they stormed with the bayonet a line of Indian artillery guns. Thomas was shot through the thigh during this battle.

41. Smythies, p346.

42. VPRS 1189 BOX 92, 55/J14458.

43. MS 6167 Box 9414 (b).

44. *Argus*, 15 December 1854.

45. Carboni, p100.

46. MS 6167 Box 9414 (b) p 54.

47. Gurry, p352.

48. Carboni, pp51, 82.

49. *Argus*, 2 October 1854.

50. VPRS 4066/P Unit 1, December 1854 no. 55.

51. *Eureka Reminiscences*, pp5, 6. It is possible that the names Brewen and Brennan are confused as misspellings of the same name by O'Brien.

52. MS 11646 Box 2178/4-5, p242.

53. MS 6167 Box 9414 (b), p55.

54. Queen v Joseph, p33.

55. Queen vs. Joseph, p24

Chapter 6

1. Pasley, MS 6167 Box 9414.

2. Anderson, Parliamentary Papers.

3. Pasley, MS 6167 Box 9414.

4. O'Brien, p89.

5. Smith, p141.

6. In Smith, pp140-141, Constable Coulson mentions that some of his fellow constables did not believe in fighting the insurgents, and that many of the soldiers were opposed to the idea, two having to be arrested for refusing orders to do so. Corfield, Wickham and Gervasoni mention that Kossack was sympathetic to the miners, as was Constable John King.

7. *Ballarat Times*: No 45: 1 January 1855.

8. *Gold Fields Commission of Enquiry Report*, pp112-3: para.2179. The testimony of police constable Robert Tully given at the State Trials confirmed this, see VPRS 6927. Annear, p84 also mentions the soldiers imbibing. This would not have been an unusual occurrence. Rum was viewed as a tonic, a necessary stimulant for the soldiers. In circumstances where the water supply was suspect at best, such as Ballarat, 'grog', in the form of rum and beer, provided soldiers with something safe to drink. The issue of grog in the British army used a time-honoured process. As each man received his one fluid ounce dram from a marked metal cup, the Company Sergeant Major marked his name off a list. When men's names could not be marked off, such as before a night mission, it was not unknown for the more enterprising soldiers to run to the back of the line for a second dram. Officers were not entitled to an issue of rum. 'Nobbler' was the colloquial expression used for a 'drink' in 1854. For a good account of how alcohol was used by British troops before and during battles, see Holmes, pp405-406.

9. MS 6167 Box 9414.

10. *Argus*, 9 December 1854.

11. Smith, p141.

12. Pasley, MS 6167 Box 9414.

13. MacFarlane, p167.

14. See Howitt, pp380–386 for some very good first hand descriptions of the Ballarat diggings. Images of the difficult terrain at Ballarat can be found in Hocking *To the Diggings*, pp46 and 90.

15. MacFarlane, p90. Corfield, Wickam and Gervasoni, p454 relate the fate of a miner named Thomas Rowlands, who tripped over a dog chain when returning to his tent one dark night and was severely injured. Rowland's fate illustrates the dangers to be faced when moving about the diggings in the dead of night.

16. Anderson, p41.

17. The moon on that night was 98% gibbous, see: http://aa.usno.navy.mil

18. The variance in estimating the time the march to Eureka began reveals the idiosyncrasies of the individual estimates of time in an era before the widespread use of personal timepieces. In evidence given during the State Trials, Lynott of the 40th guessed a time 'between three and four o'clock'. Carruthers of the mounted 40th stated 'About quarter to four'. Langley thought that the force left camp about 'Half and hour before dawn – say about half past two o'clock, or it might have been three'. Richards of the 40th said that 'We fell in at half past three o'clock, and waited till it got rather light. I think we started about five or six minutes to four'. Amos' estimate of the time was 'about three o'clock', and Webster thought they departed at 'a little after three o'clock'. In evidence taken by the Police Office in July 1855 Thomas stated that the march began at 'three o'clock'. All of these claims can be found in the records of the State Treason Trials held at the Supreme Court Library in Melbourne.

19. MS 7725 Box 646/9 Huyghue.

20. Civil twilight is the false dawn that precedes sunrise. Timings of sun and moon movements for any site on Earth back until 1700 A.D. can be found at the United States Naval Observatory site: http://aa.usno.navy.mil

21. Lynch, p31.

22. Corfield, Wickham and Gervasoni, pp183-185.

23. MacFarlane, p166.

24. Hamley, p33.

25. Smith p 141, VPRS 6927.

26. VPRS 5527/P Unit 3.

27. Information supplied by the Gold Museum, Ballarat revealed that concerns have been raised by some informed sources within the office of the Surveyor General of Victoria that there was confusion over the scales used by the cartographers when they drew the Trial Plan. In drawing the

Plan, the *Link* measurement equalling eight inches was used as the scale, when the *Foot* or twelve inch measurement should have been used. The result of this was that the route shown, while being correct in the depiction of its general direction, was truncated, and did not extend its full length, thus placing the site of the stockade much closer to Black Hill than it actually was.

28. A copy of the map can be found at the Ballarat Library. One chain equals 22 yards.

29. The Eureka Centre is located at the corner of Rodier and Eureka Streets in Ballarat Victoria. The Centre was opened in 1998, and comprises numerous galleries as well as displays depicting events surrounding the Eureka uprising. The Centre also provides a focus for the continuing debate and interpretation of Eureka.

30. *The Siting of the Eureka Stockade*, Gold Museum Ballarat, August 2004,

31. Carboni Lower Stockade, p96.

32. MS 7725 Box 646/9.

33. Turner, p71.

34. VPRS 6972.

35. Nosworthy, p24.

36. Featherstone, India, p22.

37. Adkin, p166.

38. Holmes, pp253 and 255.

39. Anderson, p42.

40. State Trail, Queen v. Joseph, p32.

41. MS 7725 Box 646/9.

42. Withers, pp123-124.

43. Queen v. Joseph, p19.

44. O'Brien, p104.

45. The original of von Guerard's painting of the Ballarat diggings hangs in the Ballarat Fine Arts museum. For the sketch of the Eureka lead refer to Bate, p28.

46. The theory that walking tracks ran close to the Eureka Stockade is the theory of Eureka and goldfields researcher Mr John Todd of Ballarat. Todd, in conjunction with Rodger Trudgeon, Manager, Curator and Deputy Museums Director at the Ballarat Gold Museum, conducted extensive research into the most likely site of the stockade. In his research Todd traced the routes of several tracks that had direct relevance to Eureka, and included these in the Ballarat Gold Museum's unpublished work *Siting the Eureka Stockade*.

47. Anne Lewis grew up on Black Hill during the 1960s. She vividly recalled playing in and among the old diggings dominating the hill in those days. One feature of those diggings mentioned by Anne was the presence of cleared tracks winding their way through the broken terrain.

48. Macfarlane, p169.

49. Queen v. Hayes, p74

Chapter 7

1. Withers, p124.

2. Queen v. Joseph, p19.

3. Corfield, Wickham and Gervasoni, p85.

4. MacFarlane, p173.

5. Ferguson, p60.

6.   Ferguson, p60.

7.   Lynch, p30.

8.   VPRS 6927.

9.   Carboni, p113, mentions some soldiers drinking brandy excessively from a pannikin dipped into a bucket and shouting obscenities as they got roaring drunk following their return from the stockade.

10. *Argus* 9 Dec, 1854.

11. VPRS 6927.

12. MS 6167 Box 9414 (6).

13. VPRS 1085/P Unit 8, Duplicate 162 Enclosure no.7.

14. These could have been the same men mentioned by Richards in VPRS 6927 as moving over a hill during the advance on the stockade.

15. Nafzinger, G.F., *Skirmisher Tactics of the Napoleonic Wars, Part 1: British Skirmishers*.

16. Queen v. Joseph, p35.

17. Lynch, p30.

18. Ferguson, p60.

19. Hocking, p193.

20. Withers, p117, also Rede's letter of 2 December, VPRS 1189 Box 92 54/J14462.

21. *Eureka Reminiscences*, p64.

22. Ferguson, *Experiences of a Forty Niner in Australia and New Zealand*, p60.

23. Monaghan, pp263-34.

24. Withers, p124.

25. Withers, p117.

26. Queen v. Joseph, p35.

27. Queen v. Joseph, p21.

28. Queen v. Joseph, p24.

29. MS 7725 Box 646/9, Huyghue.

30. *Light Magazine*, September 1977.

31. Withers, p124.

32. Corfield, Wickham and Gervasoni, p334.

33. Eureka Reminiscences, p58.

34. State Trials, Queen v. Hayes, p83.

35. MS 7725 Box 646/9.

36. Queen v. Hayes, p76.

37. Queen v. Joseph, p35.

38. Lynch, p30.

39. VPRS 1085/P Unit 8, Duplicate 162 Enclosure no.7.

40. Queen v. Hayes, p76.

41. Queen v. Joseph, p21.

42. Queen v. Hayes, p85.

43. *Argus*, 9 December, 1854.

44. Lynch, p30.

45. *Guardian* 1 Dec 2004, http://www.cpa.org.au/garchve04/1209eureka.html

46. Eureka Reminiscences, p47.

47. *Argus*, 5 December 1854.

48. Ferguson, p60.

49. Called 'Roony' by Neill.

50. *Eureka Reminiscences*, p22.

51. State Trial 1855, Queen v. Joseph.

52. A description of Juniper's leg wound, 'compound fracture of the leg due to gunshot', is in the list of casualties forwarded to the Deputy Adjutant General by Thomas 3 December. Neill in Withers, p124, states that this wound was inflicted by the insurgents' first volley.

53. Wickham, p29.

54. Anderson, p41.

55. Ferguson, p60.

56. Withers, p117.

57. Hocking, p134.

58. Corfield, Wickham and Gervasoni, p235.

59. Carboni, p64.

60. MS 7725 Box 646/9.

61. See Appendix 1 for discussion of the effects of musket fire during the era.

62. Thomas' report to Hotham, Anderson: Eureka Parliamentary Papers, p 42.

63. This was not necessarily true for post-traumatic complications of any wounds that might have developed. Wounds had to be kept clean, and infection avoided. Shreds of clothing, leather, chips of bone and particles of other foreign matter carried into the wound by the ball were common causes of infected wounds. In an era in which the medical ramifications of cleanliness and personal hygiene were only beginning to be understood, avoiding infection of wounds was often more a matter of chance than anything else.

64. Figures of 1550 feet per second velocity for musket balls, and 3250 feet per second for the M16 are given in Nosworthy, pp33-34; even so the velocity of such black-powder balls remains markedly lower than that of a modern military round.

65. *Eureka Reminiscences*, p24.

66. Magee, R., *Muskets, Musket Balls and the Wounds They Inflict*, Aust/NZ Journal of Surgery, (65), 1995 provides a very informed account of the wounds caused by musket balls.

67. *Eureka Reminiscences*, p32.

68. MS 7725 Box 646/9.

69. MS 6167 Box 9414 (b), Pasley, pp55-57.

70. VPRS 1085/P Unit 8, Duplicate 162, Enclosure no.7.

71. Lynch, p30.

72. Ferguson, p60.

73. Thomas had taken part in several actions during his service in Afghanistan and India. At Kandahar in 1842, he participated in repelling a determined assault by tribesmen who had penetrated one of the gates of the city. In 1843 he had marched with the 40th directly into the face of 23 Indian artillery guns and after a savage fight, which left more than 140 men and officers dead, captured the pieces. Thomas was wounded in the leg during this action.

74. Carboni, p64.

75. Lynch, p30.

76. MS 7725 Box 646/9, Huyghue.

77. MS 7725 Box 646/9, Huyghue.

78. O'Brien, p18,

79. Smith, pp26-59.

80. Carboni, p64.

81. Shanahan in Withers, p117.

82. Opcit, p64.

83. MS 7725 Box 646/9, Huyghue.

84. Tsouras, p92.

85. The legend that many of the Eureka insurgents were drunk when the attack on the stockade occurred has been one of the most persistent canards blighting the Eureka story. Carboni does mention a sly grog seller being ejected from the stockade on Lalor's direct orders, but makes no mention of drunkenness. Henry Nicholls, a miner who was briefly at Eureka before leaving disillusioned on the Saturday night, stated that the he found the pickets, who were Californians, drinking in the company of an attractive young lady early on the Sunday morning. He does not say that the pickets were drunk, simply implying that they were derelict in their duty at the time he found them. To provide some perspective to Nicholls' account, Molony points out that Nicholls' recollections were recorded many years after the event, that he spent only 48 hours at Eureka, and that as an opponent of the whole affair he would not be expected to give a positive slant to his descriptions of what he saw there. Fitchett in *Eureka Reminiscences*, p56, and Huyghue in O'Brien, p24, certainly claim that there was drunkenness among the defenders of the stockade. Neither Fitchett nor Huyghue were actually present at the stockade at any stage, and were relying on nothing but rumour and hearsay. Fifty years after the event, when it was acceptable, indeed desirable, to be recognised as a Eureka veteran, O'Brien, who claimed at that time to have been inside the stockade, states in *Eureka Reminiscences*, p56, that many of Eureka's insurgents were saturated with drink. A younger O'Brien, when giving evidence at the State Trials, denied ever being near the stockade during the attack. Perhaps the elder O'Brien had allowed his imaginative recollections to be influenced by the stories of others.

86. Ferguson, *Experiences of a Forty Niner in Australia and New Zealand*, p58.

87. Allan, p16.

88. Carboni, p53.

89. Carboni, p53.

90. Winders, *Mr Polk's Army*, p60.

91. Corfield, Wickham and Gervasoni, p415.

92. Corfield, Wickham and Gervasoni, p415.

93. Carboni, p64.

94. Howitt, pp381-382.

95. Lynch, p31.

96. Carboni, p97.

97. See Appendix 1.

98. Anderson, p48.

99. Anderson, p48.

100. Joseph stood trial for treason.

101. In a 4500 signature petition to Governor Hotham, Humffray and C.F. Nicholls put forward the view that granting amnesty to the American participants at Eureka and denying it to others was biased in favour of Americans.

102. Bradfield, p17.

103. Monaghan, pp205, 232, 235.

104. MS11646 Box 2178/4-5, Pierson, p237.

105. Corfield, Wickham and Gervasoni, p465.

106. Macfarlane, p9.

107. Potts and Potts, p164.

108. *Argus* 25 June 1852, Potts and Potts.

109. *Argus* 26 December 1854.

110. *The Daily Alta*, California, 20 February 1855.

111. Ironically, exactly the same narrow minded, parochial, and morally sanctimonious cultural prejudice was exhibited against Australians in California during the 1850s. Australians who had crossed the Pacific to seek gold in California were regarded with truly paranoid suspicion as vile criminals. Every crime imaginable was blamed on the 'Sydney Ducks', as Australians were called in San Francisco, until they were accused of all the misdemeanours both great and small that occurred in that city. The Vigilance Committee in San Francisco claimed that law and order could only be achieved by the Committee's extra-judicial hanging, summary deportation and refusal to allow suspect Australians to disembark at San Francisco.

112. Carboni, p94.

113. Lynch, p29.

114. *Eureka Reminiscences*, p20.

115. Historical Studies: *Eureka Supplement*, p84.

116. Greenway, p24.

117. VPRS 1085 /P Unit 8, Duplicate 162, Enclosure no. 7.

118. Anderson, p95.

119. Lynch, p30.

120. MS 7725 Box 646/9, Huyghue.

121. Carboni, p66.

122. Macfarlane, p174.

Chapter 8

1. In VPRS 1085/P Unit 8, duplicate 162 Enclosure no.7.

2. *Eureka Reminiscences*, p22.

3. *Eureka Reminiscences*, p22.

4. MS 7725 Box 646/9.

5. Corfield, Wickham and Gervasoni, p4.

6. Corfield, Wickham and Gervasoni, p390.

7. Queen v. Joseph, p26.

8. Anderson, p42.

9. O'Brien, p22. There was a suggestion that the trooper's rein may have been cut by fire from the 40th.

10. VPRS 5527/P Unit 2, Item 2: Depositions.

11. Queen v. Joseph, p20.

12. VPRS 5527/P Unit 2, Item 2.

13. VPRS 5527/P Unit 2, Item 9. Badcock probably fired at the Scots insurgent Robertson, who bore a strong resemblance to Carboni.

14. Lynch, p30.

15. Lynch, p30.

16. Further reading on the effects of flank attacks can be found in many books on the topics of the French Revolutionary and Napoleonic Wars. Equally useful are books dealing with the American Civil War. Three that one might start with are Paddy Griffith's, *The Armies of Revolutionary France*; Mark Adkin's, *The Waterloo Companion*, and Brent Nosworthy's, *The Bloody Crucible of Courage, Fighting Methods and Combat Experience of the Civil War*.

17. Carboni, p64.

18. MS 7725 Box 646/9.

19. Queen v. Joseph, p24.

20. Queen v. Hayes, p76.

21. Queen v. Joseph, p35.

22. VPRS 5527/P Unit 2, Item 9.

23. Queen v. Joseph, p24.

24. *Eureka Reminiscences*, p70.

25. Carboni, p64.

26. Carboni, p101. Carboni gave Moore, who had been shot through both legs, a drink of water.

27. O'Brien, p19.

28. VPRS 6927.

29. VPRS 5527/P Unit 2, Item 2.

30. *Eureka Reminiscences*, pp24-25.

31. *Eyewitness at Eureka*, History Today, December 2004, p34.

32. MS11646 Box 2178/4-5, p252.

33. Queen v. Joseph, p24.

34. *Argus* of 4 December 1854.

35. Queen v. Joseph, p21.

36. Withers, p124.

37. Queen v. Joseph, p20.

38. VPRS 5527/P Unit 2, Item 9, Queen v. Joseph, p20.

39. Carboni, p103.

40. VPRS 5527/P Unit 2, Item 9.

41. Ferguson, p60.

42. *Argus* 6 December 1854. The correspondent for the *Argus* conducted some 50 or so interviews immediately after the battle.

43. Boessenecker, p179.

44. So called Californians who had not been born in America.

45. Potts & Potts, p163.

46. Cusack, p60.

47. Sadleir, p58.

48. *Argus* 6 December 1854.

49. Carboni, p97, mentions an American insurgent wounded in the thigh, and later recounts on page 112 how he was arrested when treating the wounds of an American, who, as Carboni does not name him, could have been the same man.

50. *Daily Alta California*, 19 February 1855.

51. *Records of the Castlemaine Pioneers*, p155.

52. Corfield, Wickham and Gervasoni, p23.

53. MacFarlane, p104.

54. The pistol is now in the possession of the State Library of Victoria.

55. Carboni p98, Corfield, Wickham and Gervasoni, p326.

56. *Eureka Reminiscences*, p25,

57. Ferguson, p60. Corfield, Wickham and Gervasoni, p280, lists James Hull as a seaman from the steamer Oregon in the United States, who came to Victoria on the Don Juan with Ferguson. Ferguson in *Experiences of a Forty Niner in a Third of Century in the Gold Fields*, describes Hull as the engineer on the mail steamer Oregon that worked the Sacramento San Francisco run.

58. Turner, p73.

59. *Eureka Reminiscences*, p36.

60. See Queen v Hayes, pp110-112, and MS 11491 Box 33/5 Smyth's letter to W.H. Archer.

61. Sutherland in *Eureka Reminiscences*, p64.

62. *Eureka Reminiscences*, p20.

63. Bowden, p48.

64. Ferguson, p60.

65. http://www.cpa.org.au/garchve04/1209eureka.html

66. Corfield, Wickham and Gervasoni, p352.

67. Withers, p124.

68. Withers, p124.

69. Withers, p118; Carboni, p98.

70. Allan, writing in 1884 as *One of the Insurgents*, claimed that the pikemen fled before the soldiers entered the stockade. His account is at variance with those of Neill, Miller and Mrs Shanahan. Allan makes no mention of the pikemen's fight described by Neill and Miller. Perhaps by the time that fight occurred he was no longer inside the stockade himself. See Allan, p18.

71. Carboni, p101.

72. *Argus* 8 December 1854.

73. Turner, p73.

74. All these accounts can be found in Johnson; Faulds, p36; Duke, p30; Donnelly, p34; Amies, p18.

75. Corfield, Wickham and Gervasoni, p283.

76. Queens v. Hayes, p86.

77. *Eureka Reminiscences* p22

78. The King family kept the flag until 1895, when it was given to the Ballarat Fine Arts Gallery where it remains. There may have been two flags flying over the stockade during the battle. A report in the *Argus* of 4 December 1854 describes a Union Jack beneath the Southern Cross. The *Argus* of 9 December 1854 reported that, after the fall of the stockade, Constable Hugh King found a flag like a Union Jack on a prisoner. In Corfield, Wickham and Gervasoni, p357, an account is given

of Sergeant John McNeil at Spencer Street Barracks in Melbourne ripping up a flag, said to be that flown by the rebels at Eureka. If there were two flags flying at Eureka, it is conceivable that a Union Jack might have been one of them, after all the miners did claim to be defending their British rights. Perhaps it was the insurgents' Union Jack that was destroyed by John McNeil.

79. VPRS 5527/P Unit 2, Item 2.

80. VPRS 5527/P Unit 2, Item 2.

81. Queen v. Joseph, p20.

82. VPRS 5527/P Unit 2 Item 9.

83. VPRS 5527/P Unit 2 Item 9.

84. Queen v. Joseph, p21.

85. Queen v. Hayes, p76.

86. Sadleir, p63.

87. Queen v. Joseph, p22.

88. VPRS 5527/P Unit 2, Item 2, p38.

89. Sadleir, p63.

90. MacFarlane, p98.

91. VPRS 5527/P Unit 2, Item 2, pp26-27.

92. VPRS 5527/P Unit 2, Item 2, pp26-27.

93. VPRS 5527/P Unit 2, Item 2, p31.

94. VPRS 5527/P Unit 2, Item 2, p29, *Argus* 12 December 1854.

95. VPRS 5527/P Unit 2, Item 2, p20. Reed was also referred to as Read in depositions given by two policemen following the Eureka battle and Reid in newspaper transcripts of the same men's testimony.

96. *Argus* 11 December 1854.

97. *Argus* 11 December 1854.

98. *Argus* 11 December 1854.

99. *Argus* 12 December 1854.

100. Corfield, Wickham and Gervasoni, p298.

101. VPRS 5527/P Unit 2 Item 4, *Argus* 12 December 1854.

102. VPRS 5527/P Unit 2, Item 6.

103. *Argus* 11 December 1854.

104. *Argus* 11 December 1854.

105. These names appear to be misspellings of Patrick Gittens and John Hynes, both of whom were killed during the battle.

106. *Eureka Reminiscences*, p51.

107. Stoney, p125.

108. Ferguson, pp60-61.

109. Withers, p109.

110. *Argus* 9 December 1854.

111. VPRS 5527/P Unit 2, Item 2, Queen v. Joseph, p22.

112. Queen v. Joseph, p21.

113. *Argus* 4 December 1854.

114. Ferguson, p60.

115. Carboni, p98.

116. Lynch, p31.

117. MS 7725 Box 646/9, Huyghue.

118. MacFarlane, p98.

119. MacFarlane, p99.

120. Carboni, p105.

121. Corfield, Wickham and Gervasoni, pp152-153.

122. Diamond and Faulds can be found in Johnson, pp33–34 and p37.

123. Withers, p117.

124. Withers, p124. Many secondary accounts describe Hafele having the top of his skull cleaved off by Richard's sword. This story appears in *Turner's Our Own Little Rebellion*, and has been retold many times. There is, however, no primary source confirmation that any such thing happened. Neill, the only eyewitness to Hafele's death, makes no mention of the specific details. John Bird in *Eureka Reminiscences* mentions seeing the corpse of a blacksmith who had been shot in the head, and whose brains were protruding from the wound. Madden, in the same source, relates seeing the body of a blacksmith but offers no details of his death.

125. Corfield, Wickham and Gervasoni, p92.

126. MS 7725 Box 646/9.

127. Corfield, Wickham and Gervasoni, p296.

128. *Eureka Reminiscences*, p18.

129. Ferguson, p61.

130. MS 7725 Box 646/9.

Chapter 9

1. *Argus* 6 December 1854. This correspondent may have been Samuel Irwin, the correspondent for the *Geelong Advertiser*, whom Henry Turner credits with making the *man servants, maid servants, oxen and asses* reply to the loyal toast at the dinner for the US Consul on 28 November. If so, Irwin's very obvious political bias should be kept in mind when reading his descriptions of events at Eureka.

2. Turner, p94.

3. *Argus* 20 December 1854.

4. Stoney, p123.

5. Gilbert, p26.

6. MS 11646 Box 2178/4-5.

7. O'Brien, p118.

8. Hocking, *Red Ribbon* p38.

9. Claims of wanton and widespread massacre of innocent miners by Lynch, Tuohy, Alan and Shanahan are from men who, in the case of Lynch, Tuohy and Allan, were in custody. Shanahan was hiding in an outhouse, concealed from discovery, and one would presume, unable to see much of what was happening from his less than salubrious hiding place. None of these men would have been able to witness what was happening outside the stockade or beyond their immediate surrounds, and even then under conditions of emotional duress. Their recollections were many years after the event, and by the time they made their statements the gaps in their observations would have been filled in by sincere 'memories' based more on the accepted legend of the Eureka massacre rather than an objective view of what had occurred.

10. Molony, *Eureka*, p168,

11. MS 7725 Box 646/9.

12. MacFarlane, p104.

13. Anderson, p44.

14. Urban, p20.

15. Baldwin, p37.

16. Nosworthy, *Bloody Crucible of Courage*, p252.

17. Holmes, *Firing Line*, p381.

18. Gammage, p97.

19. Gammage, p259.

20. Winders, p162.

21. Evidence of the infrequency of wounds being inflicted by the bayonet in open warfare can be found in the casualty numbers from typical battles of the mid-Victorian black-powder era. In the US army's Overland Campaign fought in Northern Virginia between May and July 1864, 25,454 men were lost to gunshot wounds, while just 36 were lost to the bayonet. For a very detailed account of the losses suffered by forces in the American Civil War see Brent Nosworthy's *The Bloody Crucible of Courage*.

22. Silver, p103.

23. A good account of this battle and the other battles of the 1837 Papineau-MacKenzie rebellion in Canada can be found on the Canadian Government website - Canadian Military Heritage http://cmhg.gc.ca/cmh/en/page_425.asp

24. Featherstone, *India*, p35, Cook, p48.

25. In exactly the same manner Australian soldiers fighting in New Guinea during 1942 ensured that whatever enemies they passed over were dead. See McCarthy, p178.

26. Keegan, p127.

27. Well-drilled troops will revert to training under duress. This is the purpose of repeated practice and rehearsal. In a conversation the author had with a veteran of the French army, who fought during the Algerian conflict in the late 1950s and early 1960s, the man made the point that in combat one's training took over, and for a time normal restraints, including those of self preservation, disappeared. When the fighting was over, it required several cigarettes and a nice quiet place to sit down to cope with the realisation of what risks had been taken in the heat of battle.

28. *Eyewitness at Eureka, History Today*, December 2004m p33.

29. *Argus* 8 December 1854.

30. Winders, p137.

31. Turnbull, p133.

32. MS 6167 Box 9414 (b).

33. Grey, pp212-213.

34. MS 6167 Box 9414 (b).

35. After the fall of the Spanish town of Badajoz to the British army in 1812, many of the victorious troops went wild, and indulged in an orgy of looting and rape. Major Cameron of the 95th Rifles formed his four companies up, and threatened to shoot dead on the spot any man who broke ranks to join the affray – see Urban, p176.

36. *Argus* 4 December 1854.

37. Allan, p19.

38. To his credit, Ian MacFarlane deals with the issue in some detail in his book *Eureka From the Official Records*, pp121-128.

39. Anderson, p84.

40. MacFarlane, p124.

41. MacFarlane, p121

42. Macfarlane, p126.

43. Ferguson, p62.

44. See Urban, p176-183 for a good account of the chaos that engulfed Badajoz.

45. Baldwin, p38.

46. Edwards, p108.

47. Ferguson, p62.

48. MS 11646 Box 2178/4-5, p244.

49. MacFarlane, p125.

50. Holmes, pp310-311.

51. Allan, p18.

52. See Appendix 2.

53. *Eureka Reminiscences*, p70.

54. Tsouras, p256.

55. Henry V, Act 111, Scene 1.

56. VPRS 1085/P Unit 8, Duplicate 162 Enclosure no.17.

57. Brien and Hall are listed in WO12/5365. Brien is mentioned in AJCP M2580, Hall in Wickham, *Deaths at Eureka*, p29.

58. Corfield, Wickham and Gervsoni, p247.

59. Having to comply with the after action demands of one's political masters is nothing new for the army. Following the battle of Long Tan during the Vietnam War, the Australian government insisted that a casualty total for the enemy killed during the action be given within 24 hours. This was done, but naturally excluded many dead enemy who were discovered in the days and weeks that followed. A full account of this battle and its aftermath can be found in Lex McAuley, *The Battle of Long Tan*.

60. WO12/2971.

61. H-WAR Digest - 12 Oct 2005 to 13 Oct 2005 (#2005-188).

62. Johnson, p37.

63. Alcock, p55.

64. See Appendix 1.

65. In *David S. Terry of California Duelling Judge*, David Buchanan described how a pocket watch deflected a bullet from a pistol, even though the duel was fought at what one observer described as the 'murderously close' range of ten paces.

66. Kelly, p111, Pierson MS 11646 Box 2178/4-5, p243.

67. Corfield, Wickham and Gervasoni, p477.

68. AJCP M2579 reveals that deaths in the 40th for the years 1852, 1853, 1854, 1855 and 1856 were 17, 15, 15, 7 and 14 respectively. The death toll of 15 that occurred during the year of Eureka is consistent with all other years, with the exception of only 7 in 1855.

69. *Argus* 9 December 1854.

70. Speilvogel, p39.

71. Kelly, p111.

72. *Argus* 15 December 1854.

73. Corfield, Wickham and Gervasoni, p69.

74. Corfield, Wickham and Gervasoni give the names of 21 soldiers of the 12th who spent time in hospital following Eureka.

75. Bowden, pp22-31 provides an interesting account of the hospitals operating at Ballarat and the costs of treating patients at the time of Eureka.

76. Defeated forces in the open would normally suffer about 16 percent losses. The insurgents at Eureka suffered more than this, a result explained by the fact that they were caught in a confined space, from which escape was difficult. Losses in such circumstances can be expected to be higher. See Holmes for an excellent account of casualties suffered by the British army during the black-powder era.

77. The great shame is that many retellers of the Eureka story quote Lalor's list of dead without bothering to look beyond it. Lalor himself admitted his list did not reflect the true losses suffered by the insurgents. Mrs Shanahan's account can be found in Withers, p118.

78. *Argus* 6 December 1854.

79. *Argus* 9 December 1854.

80. VPRS 1085/P Unit 8, duplicate 162, Enclosure no.7.

81. MS 11646 Box 2178/4-5, p243.

82. Corfield, Wickham and Gervasoni, p56.

83. Corfield, Wickham and Gervasoni, p431.

84. Lynch, p31.

85. Allan, p19. This was a taunt directed at the defeated insurgents by the victorious police. 'Joe' was the derogatory nickname given to the mounted police by the miners. It was repeatedly called out as warning by the miners whenever a 'Digger Hunt' was in progress, so that miners could avoid arrest. 'Joe' derived from the name of the former governor, Joseph Latrobe. By crushing the Eureka insurgents, the police had thrown the insult back in the miners face, thus 'Joe is dead now' repeatedly

Chapter 10

1. Corfield, Wickham and Gervasoni, p95.

2. Turnbull, Clive, p46.

3. Hocking, p191.

4. http://hnn.us/artciles/1328.html

5. *Collective Memory* per Novick, Peter, as quoted in Manne, Robert, *A Turkish Tale – Gallipoli and the Armenian Genocide, The Monthly*, February, 2007, p26.

Appendix 1

1. So named after Texas Ranger Samuel. H. Walker.

2. Boessenecker, p219.

3. Boessenecker, p218 Ferguson did not die at the time, but later as his leg was being amputated.

4. http://www.twainquotes.com/Guns.html.

5. http://etext.library.adelaide.edu.au/t/twain/mark/paine/chapter11.html.

6. *Ballarat Recollections*: http://www.ballaratgenealogy.org.au/art/gay/1851.htm.

7. Boessenecker, p297.

8. The Eureka Centre at Ballarat has on display a very nice Pepperbox dug from the mud near Eureka.

9. So called after George Lovell, who was appointed Inspector General of Small Arms in 1840.

10. In 1834 the British army compared flintlock to percussion systems under controlled conditions. Of 6000 rounds fired each, 922 flintlock firings misfired, while percussion firings suffered 36 misfires. See Featherstone, D., *Weapons and Equipment of the Victorian Soldier*, p15.

11. Baldwin, p21.

12. Nosworthy, Empires p208.

13. Ritchie, p125.

14. Withers, p 118.

15. This was not necessarily true for post-traumatic complications of any wounds that might develop. Wounds had to be kept clean, and infection avoided. Shreds of clothing, leather, chips of bone and particles of other foreign matter carried into the wound by the ball were common causes of infected wounds. In an era in which the medical ramifications of cleanliness and personal hygiene were only beginning to be understood, avoiding infection of wounds was often more a matter of chance than anything else.

16. Statistics of the M16-A2 rifle can be found at http://www.hk94.com/m16-rifle.html

17. Figures of 1550 feet per second velocity for musket balls and 3250 feet per second for the M16 are given in Nosworthy, pp33-34; even so the velocity of black-powder balls remains markedly lower than that of a modern military round

Appendix 2

1. *Age*, 10 April 1855.

2. Inscribed on the monument to the fallen Eureka diggers, Old Ballarat Cemetery, Ballarat Victoria.

3. Withers, p135.

4. Keir & Lawson, pp360-61.

5. Keir & Lawson, p383.

6. Blainey, Victoria's Bloody Sunday, *Royal Auto*, November 2004, p20.